Lucio '87

Toxicology of Pesticides:
Experimental, Clinical
and Regulatory Perspectives

NATO ASI Series

Advanced Science Institutes Series

A series presenting the results of activities sponsored by the NATO Science Committee, which aims at the dissemination of advanced scientific and technological knowledge, with a view to strengthening links between scientific communities.

The Series is published by an international board of publishers in conjunction with the NATO Scientific Affairs Division

A **Life Sciences**	Plenum Publishing Corporation
B **Physics**	London and New York
C **Mathematical and** **Physical Sciences**	D. Reidel Publishing Company Dordrecht, Boston, Lancaster and Tokyo
D **Behavioural and** **Social Sciences**	Martinus Nijhoff Publishers Boston, The Hague, Dordrecht and Lancaster
E **Applied Sciences**	
F **Computer and** **Systems Sciences**	Springer-Verlag Berlin Heidelberg New York
G **Ecological Sciences**	London Paris Tokyo
H **Cell Biology**	

Series H: Cell Biology Vol. 13

Toxicology of Pesticides: Experimental, Clinical and Regulatory Perspectives

Edited by

Lucio G. Costa

Department of Environmental Health
University of Washington
Seattle, WA 98195, USA

Corrado L. Galli

Institute of Experimental Sciences
University of Milan, Via Balzaretti 9
I-20133 Milan, Italy

Sheldon D. Murphy

Department of Environmental Health
University of Washington
Seattle, WA 98195, USA

Springer-Verlag
Berlin Heidelberg New York London Paris Tokyo
Published in cooperation with NATO Scientific Affairs Division

Proceedings of the NATO Advanced Study Institute on Toxicology of Pesticides:
Experimental, Clinical and Regulatory Perspectives held at Riva del Garda, Italy, from
October 6–15, 1986

ISBN 3-540-16093-0 Springer-Verlag Berlin Heidelberg New York
ISBN 0-387-16093-0 Springer-Verlag New York Berlin Heidelberg

© Springer-Verlag Berlin Heidelberg 1987
Printed in Germany

Printing: Druckhaus Beltz, Hemsbach; Bookbinding: J. Schäffer GmbH & Co. KG, Grünstadt
2131/3140-543210

PREFACE

The protection of human health and food and fiber resources against the ravages of pests of many sorts is a continuous struggle by all people in the world. The use of chemical pesticides as an aid in this struggle is now also global. These chemicals are deliberately added to the environment for the purpose of killing or injuring some form of life. Because pesticides are generally less selectively toxic than would be desired, non-target species, including humans, must be protected from injury by these chemicals.This can only be achieved by thorough understanding of the comparative toxicology of these compounds, and by minimizing human (and other desirable species) exposure. The latter can only be achieved by sound regulatory policies that utilize scientific principles and data, properly tempered by both gaps in that data and sociologic and economic considerations.

This book contains the proceedings of the NATO Advanced Study Institute on "Toxicology of Pesticides: Experimental, Clinical and Regulatory Perspectives" held in Riva del Garda on October 6-15, 1986. This NATO-ASI has been promoted by the School of Public Health and Community Medicine, University of Washington at Seattle, by the Institute of Pharmacological Sciences, University of Milano and by the Giovanni Lorenzini Foundation, and has been sponsored by both the Society of Toxicology (USA) and the Italian Society of Toxicology.
This Institute brougth together scientists active in basic and applied research on pesticides, basic scientists who reviewed new methodologies and concepts especially relevant to pesticide toxicology, and professionals involved in the scientific aspects of policy and regulatory decisions concerning pesticides. Partecipants were from twelwe different countries and represented academic, industrial and governmental institutions.

The major thrust of this NATO-ASI was to foster discussions and exchange of perspectives on three aspects of pesticide toxicology: basic research, clinical experiences and regulatory processes.
The first eighteen chapters of this volume cover various aspects of basic toxicology of pesticides, including their effects on various organ

systems. Two chapters are devoted to the discussion of ecological problems related to the use of pesticides, while another series of papers examine clinical aspects of pesticide poisoning and other aspects of potential human exposures. Two final chapters summarize the regulatory aspects of pesticides as well as new developments in the area of pest control involving the application of biotechnology. Following the invited lectures, also included in the volume are a selected number of volunteered communications presented at the meeting, dealing with various aspects of pesticide toxicology. These presentations reflect the high degree of involvement of all participants at this conference.

Lucio G. Costa
Corrado L. Galli
Sheldon D. Murphy

Table of Contents

Toxicology of Pesticides: A Brief History

Lucio G. Costa
Department of Environmental Health, SC-34
University of Washington
Seattle, WA 98195

The Early Years

Pesticides have been used to a limited degree since ancient times. The Ebers Papyrus, written about 1550 B.C., lists preparations to expel fleas from the house. The oldest available record is Homer's mention (about 1000 B.C.) that Odysseus burned sulfur "...to purge the hall and the house and the court" (Odyssey XXII, 492-494). Pliny the Elder (23 to 79 A.D.) collected in his Natural History many anecdotes on the use of pesticides in the previous three or four centuries (Shepard, 1939). Dioscorides, a Greek physician (40 to 90 A.D.) knew of the toxic properties of sulfur and of arsenic. There are records showing that by 900 A.D. the Chinese were using arsenic sulfides to control garden insects. Veratrum album and v. nigrum, two species of false hellebore, were used by the Romans as rodenticides (Shepard, 1939). In 1669, the earliest known record of arsenic as an insecticide in the Western World mentioned its use with honey as an ant bait. Use of tobacco as contact insecticide for plant lice was mentioned later in the same century. Copper compounds were known since 1807 to have fungicidal value, and the Bordeaux mixture (hydrated lime and copper sulfate) was first used in France in 1883. Hydrocyanic acid, known to the Egyptians and the Romans as a poison, was used as a fumigant in 1877 to kill museum pests in insect collections (Shepard, 1939). Carbon disulfide has been used as an insect fumigant since 1854.

Even in this century until the mid 1930's, pesticides were mainly of natural origin or inorganic compounds. Arsenicals have played such a very relevant role in pest control that in 1939 Shepard wrote that "they will always be used in considerable quantities despite their dangerous qualities". Indeed, although the use of arsenical insecticides in agriculture has greatly decreased following the introduction of modern insecticides, the use of arsenical herbicides has increased (Hayes, 1982).

The first recorded recommendation of sulfur for the control of diseases are at the beginning of the last century (Forsyth, 1802), and around 1850 its fumigant properties were recognized. Until 1903, 95% of the world supply of sulfur came from Sicily (Shepard, 1939). Despite the introduction

NATO ASI Series, Vol. H13
Toxicology of Pesticides: Experimental, Clinical
and Regulatory Aspects. Edited by L. G. Costa et al.
© Springer-Verlag Berlin Heidelberg 1987

of organic sulfur fungicides, such as captan, maneb, and others in the late 1950's, because of the toxicological problems associated with some of them, inorganic sulfur remains one of the most important fungicides. Its major advantage is its low toxicity to man, wildlife, and the environment (Tweedy, 1981). Nicotine has been widely utilized as an insecticide all over the world, as has been rotenone, used as a fish poison in South America since 1725. Mercury chloride was extensively used as a fungicide since 1891, and was slowly replaced by its organic forms such as phenylmercury (1915), al-kyloxyalkylmercury (1920's) and alkylmercury (1940's). The latter compounds were responsible for an outbreak of poisoning in Iraq in 1971-72, due to consumption of bread made from cereal grains, and involving more than 5,000 people (Bakir et al. 1973). Several other incidents of multiple poisonings resulting from accidental ingestion of organic mercury compounds has forced stringent regulations and a decline in their use.

The first synthetic organic insecticides that appeared for public use were probably the dinitro compounds and the thiocyanates in the early 1930's. Beginning with those years, significant discoveries occurred, which led to the proliferation of new synthetic pesticides including DDT, organo-phosphates and pyrethroids. Because of their major impact on modern society, the development of these compounds will be shortly reviewed.

The Development of Modern Insecticides

The period between 1935 and 1950 was characterized by the development of DDT and other chlorinated hydrocarbon insecticides. In the late 1930's, the Swiss firm J. R. Geigy A. G. was very active in researching a contact poison for flies, mosquitoes, and other insects. Although first synthesized by Zeidler in 1874, it was not until 1939 that Dr. Paul Muller (who in 1948 was awarded the Nobel Prize in Medicine for his discoveries) found that di-chlorodiphenyltrichloroethane (DDT) acted as a contact poison on flies, mos-quitoes, and other insects. In 1940 the first patent was obtained, and from the beginning of 1942 the preparation appeared on the market under the names Gesarol and Neocid, among others (Holmstedt and Liljestrand, 1963). During World War II, the practical value of DDT was demonstrated when, with the aid of it, a severe epidemic of typhus in Naples was successfully stopped. This disease is transmitted by body lice, and after powdering 1.3 million people inside their clothing with DDT, the epidemic came rapidly and completely un-der control (Holmstedt and Liljestrand, 1963). Afterwards, one of the most

successful uses of DDT has been in the malaria eradication programs. During
the first half of this century, it is estimated that some 300 million people
suffered from malaria each year, and of those 3 million died of it. Before
the development of DDT, most of the attempts to control or eliminate malaria were unsuccessful. As an example of the effectiveness of DDT, in the
province of Latina (Italy) there were 50-60 cases of malaria per 1,000 in-
habitants in 1944. This incidence was reduced to zero by 1949, after a DDT
spraying program was started in 1945 (Hayes, 1975). Production of DDT in
the USA peaked in the early 1960's and gradually declined. In 1962 Rachel
Carson published the book Silent Spring, an impassioned denouncement of the
consequences of chemical contamination of the environment, with particular
emphasis on the bioaccumulation of DDT and its effects on bird reproduction.
This book, and the discussions which it generated, prompted the federal government to take action against water and air pollution as well as against
some persistant pesticides. As a consequence, the manufacture and sale of
DDT was banned in Sweden in 1970 and in the USA in 1973. The world market
for chlorinated hydrocarbons has remained relatively constant in the past 10
years, mainly because DDT is still used in many parts of the world for controlling disease-carrying vectors.

Table 1: Historical Development of Pesticides

Year	Pesticide
1000 BC	Sulfur is used by the Greeks
900	Arsenicals are used by the Chinese
1763	Nicotine, as crude tobacco, used as insecticide
1800's	First usage of pyrethrins in Asia
1848	First usage of rotenoids
1939	Discovery of the insecticidal properties of DDT by P. Muller
1940-50	Development of organochlorine insecticides (aldrin, dieldrin, etc.)
1944	Synthesis of parathion by G. Schrader
1950's	Development of insecticidal carbamates
1963	Chlordimeform, the first of the formamidine pesticides is synthesized by Schering, A. G.
1970's	Development of modern pyrethroids

Together with DDT, other chlorinated hydrocarbon insecticides were developed in the same years. The first one, hexachlorocyclohexane (also called benzene hexachloride, BHC) had been synthesized in 1825 by Faraday, but
its insecticidal properties were not discovered until more than 100 years
later. In the early 1940's, scientists in England and in France recognized

the gamma isomer of BHC, commonly known as lindane, as a highly potent insecticide, and introduced it to the market (Ecobichon and Joy, 1982). The cyclodiene-type chlorinated hydrocarbons were introduced to the market starting from the mid 1940's. The insecticidal properties of chlordane were described in 1945, heptachlor was introduced for agricultural use in 1948, and in the next five years dieldin and aldrin were also marketed.

In the USA the DDT share of the market has largely been taken over by organophosphorous compounds, which comprise the majority of the insecticides in current use. Their development goes back to 1932 when Willy Lange at the University of Berlin synthesized some compounds containing the P-F linkage. During the synthesis of dimethyl and diethyl phosphofluoridate, he and his graduate student G. von Kreuger, noticed the toxic effects of the vapors on themselves. They wrote "...The vapours of these compounds have a pleasant and strongly aromatic odour. But a few minutes after inhaling, a marked pressure develops in the larynx combined with breathlessness. Then, mild disturbances of consciousness set in and also a feeling of being dazzled and painful hypersensitivity of the eyes to light. The symptoms decrease only after several hours. ...Very small quantities produce the effects..." (Lange and von Kreuger, 1932). Lange seemed to be fully aware of the potentialities of organophosphorus compound as insecticides, but he soon left Germany and did not continue to work in this field (Holmstedt and Liljestrand, 1963). The father of modern organophosphorus insecticides is therefore considered Gerhard Schrader, a chemist at the I. G. Farbenindustrie (now Bayer A. G.). In the mid thirties in Germany all available resources were used by the State in building defense programs, and the importation of goods which were not considered essential was considerably reduced. The latter included nicotine and rotenone, which were the main products available for crop protection. While working in the synthesis of organic fluorine and sulfur compounds, one day in December 1936, Schrader noticed "...that, on my way home my visual acuity was somewhat reduced. By the following day vision had practically returned to normal and I resumed my work. When other visual disturbances occurred, it became quite obvious that they were caused by a new synthetic substance" (Schrader, 1959). This substance was isolated, but it was too toxic to warm-blooded animals to be used in agriculture. In the early forties, Schrader and his colleagues found a new simple method to synthetize a particular ester of pyrophosphoric acid (tetraethylpyrophosphate; TEPP) which was brought into the market in 1944 under the trade name Bladan. Interestingly, the same compound had been synthetized in 1854 by the French chemist de Clermont who also survived to report on the compound's taste.

However, almost 100 years had elapsed before its high activity as a contact insecticide was discovered. Since Bladan was not very stable in aqueous solution, new organic phosphorous compounds were later synthesized. Schrader prepared around 7,000 organophosphates by the end of the war. In 1944, a new substance (code name, E605) was synthesized, which had optimal stability and insecticidal activity. At the end of World War II, the methods for its synthesis were taken over by the Allies; later E605 was introduced into the agricultural chemicals' market under the trade name Parathion, and it subsequently became the most widely employed insecticide of this class. During those years, other compounds of much greater toxicity than parathion, such as Sarin, Soman and Tabun were synthesized as potential chemical warfare agents, but these discoveries were kept secret by the German government (Taylor, 1985). The organophosphate diisopropyl phosphorofluoridate (DFP) was first synthesized by McCombie and Saunders in 1946.

Concurrently with their synthesis, the mechanism of action of organophosphates, i.e. inhibition of acetylcholinesterase, was also discovered. German scientists during this period noted the parasympatomimetic effects of organophosphates, and found that atropine could serve as an antidote. These findings were certainly facilitated by the knowledge of the effects and of the mechanism of action of physostigmine. This alkaloid had been isolated by Jobst and Hesse in England in 1864 from Calabar beans, the seeds of Physostigma venenosum, a perennial plant in tropical West Africa, which were used for centuries as ordeal poisons in witchcraft trials (Karczmar et al. 1970). Its miotic activity and antagonism by atropine were recognized as early as 1863 and its mode of action as a cholinesterase inhibitor, resulting in accumulation of acetylcholine, was found by Loewi and Nevratil in 1926 (Casida, 1964). With this background, it is understandable that the mechanism of action of organophosphates was suggested by Gross as early as 1939. In 1949 Du Bois firmly established that parathion toxicity was due to inhibition of cholinesterase (Du Bois et al. 1949).

Other important discoveries in the history of organophosphates are the discovery of reactivation and ageing of the phosphorylated cholinesterase. In the mid 1950's Wilson found that oximes could reactivate acetylcholinesterase after inhibition by an organophosphate (Karczmar et al. 1970). This favorable development with regard to the treatment of organophosphate poisoning was somewhat counteracted by the discovery by Hobbiger (1955) that the reactivability decreases with time, i.e. the "aged" phosphorylated enzyme is not reversible by the oximes.

The capacity of certain organophosphates to induce a delayed polyneuropathy had been known since 1930, when more than 10,000 people in the U. S. were intoxicated by cresyl phosphates, used to extract ginger for illicit distilled liquor. The mechanism involved in this particular effect of organophosphates has been systematically investigated only in the last decade (Johnson, 1982; Moretto and Johnson, this volume) and is not due to inhibition of cholinesterase. The concept of metabolic conversion of organophosphates to active cholinesterase inhibitors, and that of potentiation of their toxicity by other chemicals or pesticides were two other major discoveries of the 1950's in this area (see Murphy, 1986).

Despite the early studies on physostigmine, the carbamates were introduced as insecticides only in the late 1940's and in the 1950's. The first extensively used carbamate insecticide, carbaryl, was introduced in the U.S.A. by Union Carbide in 1953.

Table 2: Some Toxicologically-related Events Involving Pesticides

Year	Event
1930's	"Ginger Jake" paralysis in the U. S. caused by cresyl phosphates
1962	Silent Spring by Rachel Carson is published
1970-73	Restriction in the use of DDT in Sweden and U. S. for its ecological effects
1971-72	Outbreak of poisoning in Iraq due to alkylmercury fungicides
1976	Poisoning of spraymen in Pakistan by malathion due to its potentiation by impurities
1977	Restriction on the use of dibromochloropropane for its toxicity on the male reproductive system
1984	Accidents in Bhopal during the manufacture of carbaryl
1986	Over 1000 tons of pesticides are spilled in the Rhine River

---

The most important group of insecticides of recent discovery is that of synthetic pyrethroids. These compounds derive from molecules originally isolated from pyrethrum flowers which were used by Caucasian tribes and in Persia since the early 1800's to control body lice. The flowers were first produced commercially in Armenia in 1828. Production started in Yugoslavia about 1840 and was centered there until World War I, then in Japan and later in East Africa (Casida, 1980). Pyrethrum extracts contain six closely related insecticidal esters, collectively referred to as the pyrethrins, whose main structures were elucidiated between 1910 and 1924 (Casida, 1980). In

the same years, hundreds of synthetic pyrethroids were synthesized, which
led to the first commercial pyrethroid, allethrin. Few new pyrethroids were
discovered following commercialization of allethrin until about twenty years
ago, when chemical studies at Sumitomo Chemical Company (Japan) and at the
laboratory of Michael Elliot (Rothamsted Experimental Station, Harpenden,
England) brought new developments to the field of pyrethroids. Some of the
most commonly used pyrethroid insecticides, such as permethrin, cypermethrin, decamethrin and fenvalerate, were synethesized in the 1970's. Because
of their low mammalian toxicity and low environmental persistence, their use
is expected to continue to increase in years to come.

Historical Development of Pesticide Regulations in the United States

As later drugs, food additives and toxic chemicals in general, the development of pesticides was paralleled or followed by a series of laws and
regulations in order to ensure their safe and effective use, and to ensure
that the health of the users and of the general population is not jeopardi-
zed.

Pesticide regulation in the United States began with the Insecticide
Act of 1910. This act prevented fraudulent efficacy claims and authorized
the seizure and banning of compounds dangerous to human health (Ross, 1980).
The increase in variety and use of synthetic pesticides in the 1940's created new problems and resulted in the passage in 1947 of the Federal Insecticide, Fungicide and Rodenticide Act (FIFRA) which deals with the marketing
aspects of pesticides. FIFRA broadened the scope of products regulated, and
established a products registration procedure administered by the U. S. Department of Agriculture (USDA). It also prohibited the distribution and
sale of chemicals in interstate commerce unless known to be safe when used
as directed and effective for the purpose shown on the label. The realiza-
tion of the potential problems associated with pesticide residues in foods,
led to amendments of the Federal Food, Drug and Cosmetic Act (FFDCA), which
covered the human health aspects of pesticides. In 1948, the Food and Drug
Administration (FDA) began establishing safe levels of residue tolerances in
foods. In 1954, the "Miller Amendment" of the FFDCA formalized the toler-
ance-setting procedure of FDA. USDA registered only pesticide uses that
resulted in no residues or in residue levels declared safe by the FDA. Data
on the efficacy of new compounds or pesticides, and on their safe levels of
residues in raw agricultural commodities had to be provided by the pesticide
industry (Ross, 1980).

In the late sixties, increased awareness of the effects of pesticides on fish, wildlife and the environment, contributed to the creation of the U. S. Environmental Protection Agency (USEPA) by Executive Order on December 2, 1970. All responsibilities for pesticide regulation were transferred to this new agency. These responsibilities included the registration of pesticides, as required under the FIFRA; the setting of tolerances as required by the "Miller Amendment" and many other research and monitoring programs related to pesticides. In 1972 legislation was enacted for the Federal Environmental Pesticide Control Act (FEPCA). The major mandate of this act was that all pesticide products previously registered had to be subjected to a review and reregistration process by EPA. This led later to the development of a process referred to as the Rebuttable Presumption Against Registration (RPAR); this is not a cancellation notice, but a notice by EPA that, on the basis of further scrutiny, potential problems exist or may exist with a certain pesticide and, unless these concerns are "rebutted" by further data, reregistration will not take place (Ross, 1980; NRC, 1980).

A short review on the regulatory status of pesticides in the USA cannot refrain from considerations of international legislation. Most developed countries have their own registration requirements and regulatory systems for pesticides and, therefore, compounds may be registered in one nation and not in another. The international trade of foods and agricultural products can pose problems with pesticides which are addressed by several international organizations. Among them are the World Health Organization (WHO) and the Food and Agricultural Organization (FAO). In 1962, these two organizations created the Codex Alimentarius Commission with the purpose of stabilizing standards for food products, which included a Codex Committee on Pesticide Residues. Some less developed countries adopt the recommendations of the WHO and FAO directly as the basis of their pesticide regulations.

Conclusion

It is clear from this short review that the last fifty years have seen tremendous progress in the development of chemical pesticides. The continued increased in the world population and the consequent need for more food supplies renders crop protection a primary area of interest and concern for legislators and scientists. In addition to efforts directed at developing new chemical pesticides that would overcome problems of tolerance and resistance, and that are more selective for the target organisms, considerable research is now being devoted to alternatives to chemical pesticides. Because of the increased public awareness of the risks associated with synthetic chemical pesticides, and coupled with the rapid advances in biotechnology, biological pest control has become a growing industry (Marquis, 1986; Fischhoff, this volume). These advances (which obviously pose new problems for regulating genetically engineered products) will most probably lead to what is known as Integrated Pest Management, an approach which will utilize chemical, biological, and systematic agricultural practices to control pests and improve crop production.

Acknowledgments

I thank Dr. Sheldon D. Murphy for critically reviewing this chapter. Grants from the National Institutes of Health (ES-03424 and OH-00054) provided financial support.

References

Bakir, F. Damluji, S.F., Amin-Zaki, L., Murtadha, M., Khalid, A., Al-Rawi, N.Y., Tikriti, S., Dhahir, H.I., Clarkson, T.W., Smith, J.C. and Doherty, R.A. Methylmercury Poisoning in Iraq. Science 181:230-241, 1973.

Casida, J.E. Esterase Inhibitors as Pesticides. Science 146:1011-1017, 1964.

Casida, J.E. Pyrethrum flowers and Pyrethroid Insecticides. Env. Health Persp. 34:189-202, 1980.

Du Bois, K.P., Doull, J., Salerno, P.R. and Coon, J.M. Studies on the Toxicity and Mechanisms of Action of P-Nitrophenyl Diethyl Thionophosphate (Parathion). J. Pharmacol. Exp. Ther. 95:79-91, 1949.

Ecobichon, D.J. and Joy, R.M. Pesticides and Neurological Diseases. CRC Press, Boca Raton, FL, 1982, pp. 281.

Forsyth, W. A Treatise on the Culture and Management of Fruit Trees. Nichols and Son, London, 1802.

Hayes, W.J. Pesticides Studied in Man. Williams and Wilkins, Baltimore, 1982, pp. 672.

Hayes, W.J. Toxicology of Pesticides. Williams and Wilkins, Baltimore, 1975, pp. 580.

Hobbiger, F.W. Effect of Nicotinhydroxamic Acid Methiodide on Human Plasma Cholinesterase Inhibited by Organophosphates Containing a Dialkylphosphate group. Brit. J. Pharmacol. 10:356-362, 1955.

Holmstedt B. and Liljestrand G. Readings in Pharmacology. Pergamon Press, Oxford, 1963, pp. 395.

Johnson, M.K. The Target for Initiation of Delayed Neurotoxicity by Organophosphorous Esters: Biochemical Studies and Toxicological Applications. IN Reviews in Biochemical Toxicology vol. 4 (E. Hodgson, J. R. Bend and R. M. Philpot, eds.), Elsevier, Amsterdam, 1982, pp. 141-212.

Karczmar, A.G., Usdin, E. and Wills, J.H. Anticholinesterase Agents. Pergamon Press, Oxford, 1970, pp. 489.

Lange W. and von Kreuger, G. Ueber Ester der Monofluorphosphorsaure. Berl. dtsch. chem. Ges. 65:1598-1601, 1932.

Marquis, J.K. Contemporary Issues in Pesticide Toxicology and Pharmacology.
    Karger, Basel, 1986, pp. 105.
Murphy, S.D. Toxic Effects of Pesticides. IN Casarett and Doull's
    Toxicology: The Basic Science of Poisons (C.D. Klaassen, M.O. Amdur and
    J. Doull, eds.) MacMillan, N.Y., 1986, pp. 519-581.
National Research Council. Regulating Pesticides. National Academy of
    Sciences 1980, pp. 288.
Ross, R.T. Statute and Legislative history of the Federal Insecticide, Fun-
    gicide, and Rodenticide Act and Its Impact on Agriculture. J. Environ.
    Sci. Health B 15:665-676, 1980.
Schrader G. Unpublished speech given to the agricultural faculty, Univer-
    sity of Bonn, December 2, 1959 (reported in Holmstedt and Liljestrand,
    op. cit., 1963).
Shepard, H.H. The chemistry and toxicology of insecticides. Burgess Publ.
    Co., Minneapolis, 1939, pp. 383.
Taylor, P. Anticholinesterase agents. IN Goodman and Gilman's, The Pharma-
    cological Basis of therapeutics (A.G. Gilman, L.S. Goodman, T. W. Rall,
    F. Murad, Eds.), MacMillan, N.Y., 1985, pp. 110-129.
Tweedy, B.G. Inorganic sulfur as a fungicide. Residue Reviews 78:43-68,
    1981.

PRODUCTION AND USE OF PESTICIDES

Marcello Lotti

Istituto di Medicina del Lavoro
Universita' degli Studi di Padova
Via Facciolati 71, 35127 Padova, Italy.

The Pesticide Manual lists 564 active ingredients in current use as pesticides (1).

The rate of introduction of new pesticides into the market is decreasing. An average of 10-15 new pesticides per year were introduced from 1955 to 1970 but less than 5 per year in the seventies. In 1956, the synthesis of 1800 chemicals led to one commercial pesticide, in 1967 the ratio was about 5000:1 and in 1976 it was already in excess of 10,000:1 (2). Nowadays it is 22,000:1 (3). Research and development expenses are in the millions of US dollars (Du Pont estimates $ 45 million for a new pesticide as compared to $3 million in the 1950s) and the time from the discovery of the pesticidal action of a chemical to its marketing is estimated to account for up to 30-50% of its patent life (3, 4). Furthermore, the risks of unexpected toxicological or ecological problems caused by a new pesticide and those associated to early insect resistance are estimated to be too high. Therefore, the strategy of the pesticide industry is at the present time, to improve uses of available pesticides by introducing different pesticides into new crops and developing new mixtures.

According to US figures, the production and the demand for pesticides declined about 20% between 1973 and 1983 to 1.1 billion lbs.(3) The primary reason for the decline of production in the pesticide

NATO ASI Series, Vol. H13
Toxicology of Pesticides: Experimental, Clinical
and Regulatory Aspects. Edited by L. G. Costa et al.
© Springer-Verlag Berlin Heidelberg 1987

industry is the increased specific activity of new generation pesticides. In fact the rate of application of insecticides has steadly declined: figures range from 1.5 kg per acre of DDT in 1942 to 20 gr decamethrin in 1975 (4) (Figure).

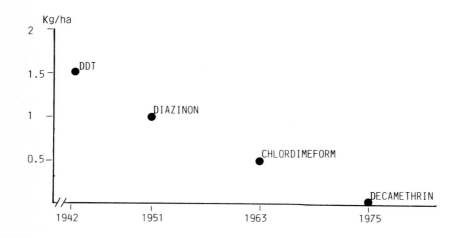

Figure 1: Evolution of rate of application of insecticides (Geissbuhler, 1980, modified)

Nevertheless the value of US sales has risen 150% over the last decade to about 4.5 billions US$, and the value per pound has almost tripled (4.00 US$ per lb), because the new generation of pesticides is higher priced (3).

Agricultural uses (68%) dominate the total pesticide consumption. Other uses include Industrial & Commercial(17%), Home & Garden (8%) and Government (7%) (3). Among reasons for a higher agricultural consumption is that pests, plant diseases and weeds cause the loss of about 30% of all crops and without the use of pesticides a further 30% would be at risk (5). This might be of vital importance in developing countries which face yet another problem - the control of vector borne diseases. These diseases, such as Malaria, Filariasis and Schistosomiasis, account for more than 650 million cases per year and they are controlled, in part, by pesticides (6).

According to their main uses and market values, pesticides can be subdivided as herbicides, insecticides, fungicides and plant growth regulators/ nematocides/ fumigants.

In the following table 1 the world agrochemical market development is reported:

	1960	1970	1980	1990*
total (US$ billions)	0.85	2.7	11.6	18.5
Insecticides	37%	37%	35%	33%
Herbicides	20%	35%	41%	39%
Fungicides	40%	22%	19%	21%
Plant Gr.Reg/others	3%	6%	5%	7%

Table 1: The world agrochemical market. *WM estimates (7)

These changes almost exclusively reflect a different use of pesticides; of interest the spectacular increase in the use of herbicides during the seventies. Of importance within the single groups a new class of pesticides, which was introduced in the late 70s, the pyrethroids. In 1983 their market value was 600 million US$ and the projected figure for 1990 is 1.9 billion US$ (7).

1990's estimates of the sales of pesticides in different areas indicate more than 55% in USA and Western Europe alone (8) (table 2).

Area	% of pesticide production
USA	31.0
W. EUROPE	25.0
E. EUROPE	10.3
JAPAN	10.3
REST OF THE WORLD	23.4

Table 2: Geographical differences in sales of pesticides: 1990 estimates (GIFAP 1983)

Western Europe and North America use 70% of the herbicide production, while developing countries use 62% of the insecticides and Western Europe alone 43% of the fungicides. These differences in the pattern of uses of pesticides reflect different tecnologies (herbicides), geographical positions (insecticides) and target crops (fungicides) (7).

16 chemical companies accounts for the 80% of the world market of pesticides. Bayer and Ciba- Geigy alone represent the 25-30%. (8)

It is clear that pesticide use is unlikely to change in the near future as their considerable benefits to agriculture and public health programmes ensures their continued usage. The scale of pesticide use is however cause of concern (2), and even though no clear association has emerged between increasing use of pesticides and incidence of chronic diseases (9), there is the possibility of unforseen and unforseeable effects on man. Moreover it should be noted that no comprehensive information exists on the exposure of the world population to pesticides, in the occupational or the general environment (10).

It is also increasingly being recognized that this extensive use of pesticides is not always correct (5). Several cases of accidental acute poisoning occour every year, particularly in the developing countries (11). The global estimate of the extent of unintentional acute pesticides poisoning vary according to different reporting criteria (5) (Table 3).

CASES	DEATHS	CRITERIA
1,111,000	20,000	From area mortality reports
873,000	16,000	From area morbidity surveys
1,057,000	20,000	National mortality data from all accidental poisonings
834,000	3,000	Mortality statistics from acute pesticide poisoning

Table 3: Global estimates of the extent of unintentional acute pesticide poisoning. From data submitted to WHO (WHO 1986).

Figures change dramatically among different countries. For instance in California, USA (population= 23,000,000) the illnesses related to pesticide exposure were 1,355 in 1985, with two deaths (12). In Sri Lanka (population= 14,000,000) data from 1975 to 1980 indicate about 13,000 hospital admission per year because of pesticides poisoning and about 1,000 deaths (13).

Table 4 compares cases of pesticide poisoning in USA and Egypt. Absolute numbers are obviously not comparable but the trends in both countries over a period of 10 years seem to be identical i.e. the number of cases and deaths is fairly constant.

YEARS	NUMBER OF CASES / NUMBER OF DEATHS IN	
	EGYPT	USA
1966-68	3969 / 113	
1968		4109 / 72
1969-71	3603 / 132	
1970		5299 / 44
1972-74	3753 / 79	
1975		6271 / 30
1975-77	4184 /97	
1976		5730 /31

Table 4: Pesticide poisoning. Comparison between Egypt and USA (1966-1977) (El Gamal 1983; ACS, Washington D.C. 1978)

Is this the unavoidable toll to be paid for the use of pesticides? The adoption of an international code of conduct on distribution and use of pesticides (14) might in part ameliorate these worrisome figures.

REFERENCES

1. The Pesticide Manual. A World Compendium. (Worhting C.R. Editor) The British Crop Protection Council, 1983.
2. Royal Commission on Environmental Pollution. Seventh Report. Agriculture and Pollution. Her Majesty's Stationery Office. London 1979.
3. Storck W.J. Pesticides Head for Recovery. C&EN, April 9, 35-57, 1984.
4. Geissbuehler H. How safe are pesticides? Experientia, 38, 890-895, 1982.
5. World Health Organization. Planning strategy for the prevention of pesticide poisoning. WHO/VBC 86.926, 1986.
6.Jeyaratnam J. Health problems of pesticide usage in the Third World. Brit. J. Ind. Med. 42, 505-506, 1985.
7. Wood, Mackenzie & Co. Agrochemical Service, Edinburgh and London 1984.
8. GIFAP Bulletin , September 1983.
9. Hayes W.J.Jr. Pesticides and human Toxicity. Ann. N.Y. Acad. Sci. 160, 40-54, 1969.

10. IARC Monographs on the evaluation of the carcinogenic risk of chemicals to humans. Miscellaneous pesticides. Vol. 30. Lyon, 1983.
11. Bouguerra M.L. Les poisons du tiers monde. Editions La Decouverte, Paris, 1985.
12. California Department of Food and Agriculture. Summary of reports from physicians of illnesses that were possibly related to pesticide exposure during the period January 1- December 31, 1985 in California. Sacramento CA, HS-1370, February 18, 1986.
13. Jeyaratnam J., De Alwis Senewiratne RS, Copplestone JF. Survey of pesticide poisoning in Sri Lanka. Bull. WHO 60, 615-619, 1982.
14. The Resolution of the 23rd session of the FAO Conference of the International Code of Conduct on the Distribution and Use of Pesticides, FAO, Rome, 1985.

Role of Metabolism in Pesticide Selectivity and Toxicity

S. D. Murphy, Ph.D.
Department of Environmental Health, SC-34
School of Public Health and Community Medicine
University of Washington
Seattle, WA, 98195

INTRODUCTION

Ideally, pesticides will be selectively toxic to target organisms and non-toxic to non-target or desirable organisms. Theoretically, this might be achieved by several means. First, a pesticidal chemical might be designed or selected which uniquely attacks a functional or structural bio-system that is peculiar to the pest organism and which is either absent in or less critical to the desirable organism. Examples of this would include: the use of red squill as a rodenticide, particularly useful because rats do not effectively vomit the material whereas other mammals do; or the use of chitin synthetase inhibitors which are selectively toxic to invertebrates with exoskeleton and nontoxic, or generally much less toxic, to mammalian organisms; or the use of juvenile hormones which affect the peculiar metamorphic development of insects.

A second approach to achieve selectivity, somewhat related to the first, would be to select or design compounds which are reactive with specific target molecules in pest species and much less reactive with target molecules in nonpest species. Although absolute selectivity would not be achieved so long as both types of organisms contained the target molecules, this principle is used to advantage in some cases in which, for example, insect species' cholinesterase is particularly more sensitive to inhibition by an anticholinesterase compound than is mammalian cholinesterase. Finally, exploitation of differences in biotransformation rates is a third mechanism of achieving selectivity in the toxic action of chemicals. Thus it may be possible to design or select compounds for pesticides which are metabolically activated more rapidly (or to a larger extent) in pest species than in nonpest species: or, alternatively, are metabolically detoxified by the nonpest species more rapidly than by the pest species. This mechanism of selectivity in toxicity is the primary focus of this paper.

Interest and concern for selectivity of toxicity extends beyond merely a consideration of target and nontarget species for the pesticides. It also is involved in understanding : (1) the potential variation in hazard to a variety of nontarget species within the environment, (2) the

NATO ASI Series, Vol. H13
Toxicology of Pesticides: Experimental, Clinical
and Regulatory Aspects. Edited by L. G. Costa et al.
© Springer-Verlag Berlin Heidelberg 1987

and (3) the environmental and genetic factors that may influence the susceptibility of individuals with in any given species. In short, understanding the metabolic basis of pesticide selectivity in toxicity may allow us to more effectively predict and control : (1) relative hazards to different desirable species in the environment, including man, (2) different susceptibilities to the hazards among individuals within the same species, and (3) the possibility of toxicologic interactions between the pesticide and other chemicals to which organisms may be exposed.

Historically, metabolism/toxicity studies on organophosphorus insecticides were among the earliest research that led to the concept that chemicals can be made more toxic as a result of endogenous metabolism. Studies. In the early 1950's (DuBois, etal, 1950; Diggle and Gage, 1951; Myers, etal, 1952) demonstrated the hepatic oxidative activation of parathion and OMPA. The further development of the concept that phosphorothionate and phosphoramidate insecticides could be made more active by metabolism was aided by the knowledge that the acute toxicity of the organophosphorus compounds was inhibition of acetycholinesterase in nerve tissue (DuBois and Mangun, 1947). This allowed the use of this specific target enzyme to study in vitro reactivity of metabolites with the critical biological macromolecule, even in the absence of chemical identification of metabolites. These early studies showed that the insecticides parathion and OMPA were not potent cholinesterase inhibitors in vitro but had to be metabolized by an oxidative system, then demonstrated by the use of liver slices, to potent anticholinesterase derivatives. Within a few years, these and other investigators (Davison, 1955; Murphy and DuBois, 1956,1957) demonstrated that pyridine nucleotide cofactor-dependent microsomal enzymes oxidatively converted organophosphorothionate insecticides to their anticholinesterase phosphate derivatives. Thus the toxicology of the phosphorothionate insecticides was intimately linked with the early developments in research on the oxidative metabolism of xenobiotics in general(Brodie, etal, 1955), and was one of the earliest specific demonstrations that foreign chemicals could be made more biologically active by enzymes in living organisms. Until that time, metabolism of chemicals had been generally associated with their detoxification.

A consideration of the role of metabolism of complex organic pesticide molecules in relation to their selective toxicity involves the broad principles of xenobiotic metabolism in general. Biotransformation of organic compounds is generally classified as two types: the so-called Phase 1 reactions, namely oxidation, reduction, or hydrolysis; and Phase 2 reactions which usually include the various conjugation systems. Frequently a single compound will undergo multiple Phase 1 and Phase 2 reactions and the ultimate toxicity will be dependent upon the balance among the rates of biotransformation through different pathways as well as the rates in specific critical target tissues. In other words, the ultimate toxicity will be dependent upon the amount and the rate at which the active metabolic derivative or the original active parent insecticide reaches the critical target sites for toxicity.

ORGANOPHOSPHATE METABOLISM-TOXICITY RELATIONSHIPS

These principles can be illustrated with a single common pesticide, methyl parathion, as shown in figure 1.

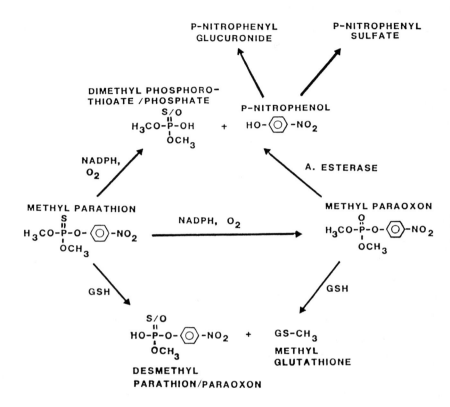

Figure 1. Biotransformations of Methyl Parathion by phase 1 and phase 2 metabolic reactions.

All of the metabolic reactions shown here for methyl parathion are catalyzed by enzymes present in mammalian liver cells. Thus, an oxidative reaction leads to the formation of the oxygen analog, methyl paraoxon, the active metabolite that inhibits acetycholinesterase. Oxidative cleavage to dimethyl phosphorothioate and p-nitrophenol, hydrolysis of methyl paraoxon and glutathione-linked demethylation of the parent compound or its oxygen analog are enzyme-catalyzed reactions that yield relatively non-toxic derivatives. Thus, for methyl parathion both activation and a detoxication reactions can be classified as phase 1. This includes oxidative desulfuration and oxidative cleavage, and hydrolysis of methyl paraoxon into its relatively nontoxic derivatives dimethylphosphate and p-nitrophenol. There is also the possibility for a variety of conjugation reactions. For example, the derivative p-nitrophenol can be conjugated either as the sulfate or the glucuronide. This may be an important consideration in biological monitoring of exposure. If one measures p-nitrophenol in the urine as an index of exposure, it is important to recognize that this may also exist as conjugates. Removal of a methyl group, catalyzed by glutathione alkyl transferase, might be viewed as a type of conjugation reaction, since the methyl group of methyl parathion becomes attached to glutathione. However, the effect is more like a phase 1 reaction to yield the more polar, less toxic desmethyl parathion or desmethyl paraoxon.

The literature is replete with studies of the metabolism of phosphorothioates, phosphorothionates and other pesticides which demonstrate production of multiple metabolites both in vivo and in vitro. Less effort has been dedicated to attempts to correlate these metabolic pathways with the toxicity of the parent compounds. One approach to this kind of question is to attempt to block (or otherwise reduce the activity of) one or more of the metabolic pathways and to observe, under these conditions, whether there has been a change in the susceptibility of the organism to poisoning by the parent compound. A series of such studies was the topic of research by students and fellows in my laboratory several years ago. (Kamienski and Murphy, 1971; Mirer, etal, 1977; Levine and Murphy, 1977a, 1977b).

It is well established that the insecticide synergist piperonyl butoxide can inhibit oxidative reactions and in some cases this can explain its insecticidal synergism with various pesticides, especially pyrethrins (Casida, etal, 1983). When we first considered this from the standpoint of mammalian toxicity, there had been little or no investigation of the potential interaction of piperonyl butoxide, which is an inhibitor of mixed function oxidases, on the mammalian toxicity of some common phosphorothionate insecticides. We undertook a series of studies in which we pretreated animals with mixed function oxidase-inhibiting doses of piperonyl butoxide (as well as another inhibitor, SKF-525A), and measured the susceptibility of these mice to challenge doses of methyl- and ethyl- parathion as well as to azinphos methyl and azinphos ethyl (ie, methyl- and ethyl-Guthion). As shown in figure 1 for methyl parathion, a mixed function oxidase enzyme system is responsible for the metabolic activation of these insecticides, converting them from the thiono phosphate to the phosphate derivatives. Detoxification of parathion by oxidative cleavage to diethylthiophosphate and p-nitrophenol had also been demonstrated to occur directly (Neal and DuBois, 1965). Therefore, the effect that mixed function oxidase inhibitors would have on the phosphorothionates' toxicities was difficult to predict. To our surprise, there was a striking difference in the effect of mixed function oxidase inhibitors on the toxicity of the methyl and the ethyl homologs of these 2 compounds (table 1).

Table 1: EFFECT OF PIPERONYL BUTOXIDE AND SKF-525A ON DIMETYL AND DIETHYL PHOSPHOROTHIONATE TOXICITY IN MICE

24 Hour LD_{50} (mg/kg)[a]

Insecticide Challenge	Corn Oil (1 ml/kg)	Piperonyl Butoxide (400 mg/kg)	SKF-525A (50 mg/kg)
		Pretreatment	
Methyl Parathion	7.6	330	220
Ethyl Parathion	10.0	5.5	6.1
Methyl Guthion	6.2	19.5	11.8
Ethyl Guthion	22.0	3.4	9.1

[a]Mice were given pertreatment compounds, i.p. 1 hour before injection, (i.p.) with graded doses of the challenge insecticides. Data adapted from Levine and Murphy (1977a, 1977b).

The toxicities of the ethyl substituted compounds were relatively little affected (some potentiation) by the pretreatment whereas the toxicity of the methyl derivatives, particularly methyl parathion, was markedly reduced by inhibition of mixed function oxidase enzyme activity.

Subsequently, we conducted a series of metabolisms studies with methyl- and ethyl-parathion which we feel established the mechanism for this marked differential toxicity interaction (Levine and Murphy, 1977a, 1977b; Murphy, 1980).

In brief, it was determined that the mixed function oxidase inhibitors inhibited both the oxidative activation, that is the production of the oxygen analog, as well as the oxidative cleavage of the parent phosphorothionates(see figure 1, for example). In the case of parathion, this resulted in a delayed onset of toxic signs which corresponded to a delayed inhibition of nerve cholinesterase. Ultimately, however, sufficient quantity of paraoxon was produced to react with sufficient acetyl cholinesterase to produce death, even at lower doses of parathion than for control animals that were not pretreated with piperonyl butoxide. This could explain the approximate 2-fold potentiation of parathion. The only major metabolic difference between methyl parathion and parathion is that both methyl parathion and methyl paraoxon can be detoxified by an alternate, non-oxidative detoxification system, namely glutathione transferase. This system is not an effective detoxification pathway for the diethyl homologues, parathion and paraoxon, however. Thus, with a delayed activation of methyl parathion due to inhibition of mixed function oxidases, there was opportunity for glutathione transferase to dealkylate methyl parathion and thus reduce the concentration sufficiently so that insufficient methyl paraoxon could be formed to cause lethal inhibition of acetylcholinesterase.

This scheme appeared to offer a plausible metabolic explanation of the marked protective effect of piperonyl butoxide against methyl parathion. However, we conducted an additonal series of tests in which we measured the rate of spontaneous in vitro reactivation of inhibited cholinesterase in tissues of mice which were given equi-inhibiting doses of methyl parathion or ethyl parathion in vivo (Levine and Murphy, 1977a). We found that the reversibility of the inhibition of brain cholinsterase produced by the ethyl derivative was very slow, whereas it was fairly rapid with the methyl phosphorylated cholinesterase. Thus, taken together, a delayed rate of activation to the oxygen analog, an alternative pathway for detoxication of the parent thiophosphate and a relatively rapid rate of reversal of inhibited acetylcholinesterase (that was slowly formed because of the delayed activation) seemed adequate to account for the 30-40-fold protection of mixed-function oxidase inhibitors against methyl parathion toxicity.

These studies illustrate several important points with respect to attempting to correlate metabolism with selectivity or differential toxicity. Firstly, they show that close structural similarity among compounds is not necessarily an assurance of close similarity in interactions with other chemicals. Secondly, they illustrate that when there are alternate pathways to achieve the same end, namely detoxication, the toxicologic influence of an inhibitor of a single pathway may be quite different from total inhibition of detoxication. Thirdly, they illustrate that both metabolic and target enzyme kinetics may interact differently with closely related compounds to produce unexpected or unexplained selective toxicity or toxic interactions.

During our studies of this interaction with methyl parathion, we determined that depletion of glutathione with diethylmaleate potentiated methyl parathion toxicity in mice by about eight-fold (Mirer, etal., 1977). This significant potentiation due to inhibition of one of the enzymes of detoxication, it is similar to the potentiation of fenitrothion that had been observed in mice that had been depleted of glutathione by pretreatment with methyl iodide (Hollingworth, 1970). These and other selective inhibition studies have substantiated that glutathione alkyl transferase, and to some extent, glutathione aryl transferase, can represent important pathways of detoxication for phosphorothionate insecticides. Alterations in the concentrations and activities of these transferase enzymes may also contribute to selective toxicity among different species or different individuals of the same species.

Several studies have demonstrated that the oxidative metabolism of the phosphorothionate insecticides can lead to either more or less toxic metabolites. For several years an apparent anomaly existed in which experimentally demonstrated variations in the in vitro metabolism of phosphorothionates to their oxygen analogs by liver preparations, was inconsistent with what would be predicted for the same variables with respect to alterations in susceptibility to poisoning (Murphy, 1969; DuBois, 1969; Alary and Brodeur, 1969). For example, when strictly the activation reaction was examined, several investigations showed that male rat livers or phenobarbital-induced rat livers produced a greater quantity of the oxygen analog than female rat livers or uninduced animals respectively. In contrast, male rats or rats that had been induced by phenobarbital, were more resistant to poisoning by the phosphorothionates than female rats or uninduced rats, respectively. When it was recognized that the liver also contained a mixed function oxidase detoxication system which responded to inducers or hormonal affects similarly to the activation system (Neal, 1967; Nakatsugawa and Dahm, 1967), it was easier to explain the apparent discrepancy between liver metabolism and susceptibility to poisoning. It also led to studies that indicated that, although the liver was the most active tissue with regard to oxidative biotransformation of these compounds, other extra hepatic tissues were also capable of activating the phosphorthionates (Poore and Neal, 1972). Of particular significance was activation in target tissues, such as brain or lung, where cholinergic nerves exert critical control. It appeared therefore, that the liver probably represented predominately a detoxifying system. The search for correlation between metabolic activation and toxicity still needs more focus on metabolism in extra-hepatic target tissues (Murphy, 1972a; Casida, 1983).. Although there has been some research in this area, the relatively low activity of the bioactivation enzymes in the target tissues has made progress in establishing in vitro/in vivo correlations for activation less rapid than for detoxification systems.

METABOLIC VS TARGET ENZYMES

It is also apparent that both metabolic and target enzymes are important in establishing biochemical predictors for differences in susceptibility to parathion and other organophosphate insecticides among non-target species. Several years ago, review of the comparative toxicity literature, revealed some rather striking differences in the sensitivity of avian, mammalian, and fish species to the toxicity of parathion, and some closely related compounds (Murphy, 1972b). Initially, it was hypothesized that this would be explainable on the basis of differential metabolism. However, measurement of the activity of critical activation and inactivation pathways in livers of representative species from these three classes of animals, failed to explain the differential toxicity of the compounds (Murphy, 1966, 1972b). Further studies indicated that differences in the sensitivity of the target enzyme, cholinesterase, in the nerve tissues of the different species was undoubtedly, a major factor in determining species susceptibility to poisoning with some compounds (Murphy, etal, 1968; Wang and Murphy, 1982). Detailed knowledge of comparative kinetics of the interaction of insecticides with both target and biotransformation systems may enable development of in vitro or quick test systems to screen compounds which are selectively more or less toxic for a given class of animals (Murphy, etal, 1984).

ENZYMATIC HYDROLYSIS AND TOXICITY

Hydrolysis is a "Phase One" reaction that has been demonstrated to be critical in determining some aspects of pesticide selectivity and toxicity. An extremely important detoxication reaction for many organophosphates is the hydrolysis of the P-O-C bond catalyzed by so-called A-esterases (Murphy, etal, 1976). Hydrolysis of certain carboxyl or carboxy-amide linkages in some phosphorothionate or phosphate insecticides, and hydrolysis of ester or amide linkages in carbamate insecticides and herbicides can also represent important detoxication pathways (Murphy, 1969; Su, etal, 1971; Singleton and Murphy, 1973; Murphy, etal, 1976). In some cases the selective toxicity to target organisms takes advantage of deficiencies in hydrolytic detoxification systems (Krueger, etal, 1960), and there are several examples of strains of insects whose insecticide resistance appears to be dependent upon the fact that they have developed a high titer of the hydrolytic enzymes that detoxify insecticides (Matsumura and Brown, 1963; Glickman and Casida, 1982). Similarly, the resistance of desirable rice plants, as compared to other weeds and grasses, to injury by the herbicide propanil is based upon the fact that the rice plant contains an amidase which rapidly hydrolyzes propanil, whereas the target weeds and grasses do not contain this system (Matsunaka, 1968). In laboratory animals, the activity of the amidase which hydrolyzes propanil is important in determining the quality of the toxic effects of this herbicide Singleton and Murphy, 1973). Thus in animals in which normal amidase activity existed, administration of propanil caused cyanosis due to the methemoglobinemia produced by the aniline derivative that was released upon hydrolysis of the herbicide. On the other hand, when the amidases were inhibited by organophosphate compounds, less methemoglobin was produced, and the quality of the signs and symptoms of propanil poisoning changed to more generalized nonspecific depression.

Perhaps the best known and most striking case of esterase determination of selective toxicity is that of malathion. Mammals have a high titer of malathion esterase which hydrolyzes the carboxyl ester groups of malathion (or its oxygen analog) and thereby rapidly detoxify it (Krueger and O'Brien, 1959). This can explain the relatively low mammalian toxicity of malathion. Malathion-sensitive insects, however, had relative low titers of malathion esterase activity in their tissues.

Several years ago, it became well established that the carboxylesterase that hydrolyzes malathion and related carboxyester-containing insecticides is sensitive to inhibition by several other organophosphate insecticides, at otherwise nontoxic doses. When animals were exposed to carboxylesterase-inhibiting compounds they became, under some circumstances, as much as 100-fold more susceptible to the acute toxicity of malathion than control animals with normal carboxylesterase activity (Frawley, etal, 1957; Murphy, etal, 1959). Initially, these observations led to a regulation by the US Food and Drug Administration, that each newly introduced anticholinesterase insecticide had to be tested in combination with all other approved anticholinesterases to determine if potentiation was likely. Ultimately it was determined that under chronic feeding conditions at or close to the levels that might be expected to be ingested in the diet, the hydrolytic metabolism appeared not to be a rate-limiting factor in organophosphate toxicity. Therefore, toxicological interactions seemed less likely and the testing requirement was discounted.

Nevertheless, an important principle was illustrated by the experimental demonstration of malathion potentiation. That is, the utilization of nontarget species' metabolic detoxication systems as the basis for selective toxicity has some attendant risks. If environmental or genetic factors greatly alter this detoxication system, and marked susceptibility to poisoning, by what was considered a rather nontoxic material, might occur. In fact, this phenomena appears to account, at least in part, for an incident of mass poisonings, by malathion in a malaria control spray program in Pakistan (Baker, etal, 1978). In that incident, several thousand spray workers became ill, and several died, while working with malathion, one of the insecticides used in the mosquito control program. Investigation showed that workers with opportunities for exposure and varying degrees of blood cholinesterase inhibition depending upon the particular sample of technical-grade malathion that was being used. Further investigation showed that the sample of malathion that was responsible for most of the intoxications contained a greater number and higher concentrations of contaminants which were relatively potent inhibitors of malathion esterase (Miles, etal, 1979; Talcott, etal, 1979). This incident of occupational poisonings appears to be explained by synergism of an active ingredient of an insecticide by contaminants or minor ingredients of the technical product.

ACUTE VERSUS CHRONIC TOXICITY

The insecticide DDT may be cited as an example of how biotransformation may have different acute versus chronic toxicity outcomes (Murphy, 1986). DDT undergoes metabolic transformations leading ultimately to DDE or DAA. These two end products of two different pathways have acute toxicities that are less than 1/4 the toxicity of the original compound in laboratory rats. Metabolism to DDA is clearly a detoxification pathway and leads to excretion. However, the toxicological importance of metabolism to DDE, is not so clear cut. DDE is long stored in body fat, and it is equally or more active than DDT as an inducer of microsomal endoplasmic reticulum of the liver. It is also alleged to have some hepato-tumorogenic effects, and thus, may not represent a detoxication pathway with respect to chronic toxicity. Most studies of metabolism toxicity relationships of pesticides have focused on acutely toxic actions. Unless these acute events accurately reflect chronic outcomes as well, (ie, they are the initiating event of chronic sequelae), the role of metabolism may be misinterpreted from the broad toxicologic viewpoint.

The mutagenic activity of chemicals in bacterial organisms, although short of a specific molecular target, has frequently been used to represent evidence of potential for yielding the chronic end point of carcinogenicity. Shuphan, etal, (1981), studied dialate and trialate herbicide metabolism with respect to the mutagenic activity of the metabolites. They showed that oxidation and other pathways of metabolism of these two herbicides catalyzed by mouse liver microsomal systems resulted in the formation of relatively potent mutagenic metabolites. These herbicides have been reported to be carcinogenic in mice and rats. It appears therefore, that studies of the metabolic routes that lead to mutagenic activities of pesticides in simple in vitro test systems may be a useful approach to attempt to develop a body of knowledge that would relate metabolic transformation characteristics to chronic carcinogenic activity.

Studies by Bleeke, etal. (1985) examined the metabolism of metribuzin herbicide with respect to its hepatotoxicity. These investigators found that pretreatment with the mixed function oxidase inhibitor piperonyl butoxide reduced the hepatotoxicity and the lethality of metribuzin by three-fold. Metribuzin also depleted the livers of treated animals of their glutathione content. When animals were administered diethylmaleate to deplete liver glutathione content, they became twice as susceptible to poisoning by metribuzin as unpretreated animals. The metabolic explanation of these observations is that metribuzin is oxidized by mixed function oxidases to a more active derivative, namely, the sulfoxide which can interact with a critical liver protein. This reaction appears to be responsible for its hepatotoxicity. Piperonyl butoxide blocks this oxidation and also protects against the toxicity (by three-fold). In addition to covently binding with liver protein, the sulfoxide derivatives of metribuzin will conjugate with glutathione. Glutathione, then, can act as a sparing molecule to protect critical proteins from the activated metabolite. When the sparing action of glutathione is reduced by compounds that deplete glutathione, the toxicity of metribuzin is potentiated.

SUMMARY

The foregoing discussion illustrates a number of examples of metabolic transformations of pesticides that are known to influence their toxicity to either target or nontarget species. Knowledge of relationships of this kind are absolutely essential for understanding and predicting situations of chemical interactions that result in either increased or decreased toxicity, for assessing the toxicologic significance of genetic variability in biotransformation pathways, and for evaluating broadly the relative hazard to various nontarget species that coinhabit our environment. Acquisition of this information is essential for the safe use and improved design of these useful biologically active materials.

ACKNOWLEDGEMENT

The author's past and current research on pesticide metabolism-toxicity relationships is supported in part by grants from the National Instituteof Environmental Health Sciences (ES-03424) and the Charles A. Dana Foundation.

REFERENCES

Alary, K. P. and Brodeur, J. Studies on the mechanism of phenobarbital-induced protection against parathion in adult female rats. J Pharmacol Exp Therap 169:159-167, (1969).

Baker, E.L., Zack, M., Miles, J.W., Alderman, L., Warren, M., Dobbin, R.D., Miller, S. and Teeters, W.R. Epidemic malathion poisoning in Pakistan malaria workers. Lancet 1:31-34, (1978).

Bleeke, M.S., Smith, M.T., and Casida, J.E. Metabolism and toxicity of metribuzin in mouse liver. Pest Biochem Physiol 22:123-130, (1985).

Brodie, B.B., Axelrod, J., Cooper, J.R., Gaudette, L., LaDu, B.N., Mitoma, C., and Undenfriend, S. Detoxification of drugs and other foreign compounds by liver microsomes. Science 121:603-604, (1955).

Casida, J.E. Propesticides:bioactivation in pesticide design and toxicological evaluation in Pesticide Chemistry:Human Welfare and the Environment, Vol. 3, J. Miyamoto etal, eds. Pergamon Press, Oxford, pp 239-246, (1983).

Casida, J.E., Gammon, D.W., Glickman, A.H., and Lawrence, L.J. Mechanisms of selective action of pyrethroid insecticides. Ann Rev Pharmacol Toxicol 23:413-438, (1983).

Davison, A.N., The conversion of schradan (OMPA) and Parathion into inhibitors of cholinesterase by mammalian liver. Biochem J. 61:203-209, (1955)

Diggle, W.W., and Gage, J.C. Cholinesterase inhibition by parathion in vivo. Nature 168:998, (1951).

DuBois, K.P. Combined effects of pesticides. Can Med Assoc J 100:173-179, (1969).

DuBois, K.P., Doull, J., and Coon, J.W., (1950). Studies on the toxicity and pharmacologic action of octamethyl pyrophosphoramide (OMPA;Pestox III). J Pharmacol Exper Therap 99:376-393.

DuBois, K.P. and Mangun, G.H. Effect of hexaethyl tetraphosphate on choline esterase in vitro and in vivo. Proc Soc Exper Biol Med 64:137-139, (1947).

Frawley, J.P., Fuyat, H.N., Hagen, E.C., Blake, J.R., and Fitzhugh, O.G. Marked potentiation in mammalian toxicity from simultaneous administration of two anticholinesterase compounds. J Pharmacol Exper Therap 121:96-106, (1957).

Glickman, A.H. and Casida, J.E. Species and structural variations affecting pyrethroid neurotoxicity. Neurobehav Toxicol and Teratol 4:793-799, (1982).

Hollingworth, R.M. Dealkylation of organophosphorus triester by liver enzymes. In : Biochemical Toxicology of Insecticides, R.D. O'Brien and I Yamamoto, eds. New York: Academic Press, Inc., New York, 1970, pp75-92.

Kamienski, F.X., and Murphy, S.D. Biphasic effects of methylenedioxyphenyl synergists on the action of hexobarbital and organophosphate insecticides in mice. Toxicol Appl Pharmacol 18:883-894, (1971).

Krueger, H.R., and O'Brien, R.D. Relationships between metabolism and differential toxicity of malathion in insects and mice. J Econ Entomol 52:1063-1067, (1959).

Krueger, H.R., O'Brien, R.D., and Dauterman, W.C. Relationship between metabolism and differential toxicity in insects and mice of diazinon, dimethoate, parathion and acethion. Econ Entomol 53:25-31, (1960).

Levine, B.S., and Murphy, S.D. Esterase inhibition and reactivation in relation to piperonyl butoxide-phosphorothionate interactions. Toxicol Appl Pharmacol 40 379-391, (1977a).

Levine, B.S., and Murphy, S.D. Effect of piperonyl butoxide on the metabolism of dimethyl and diethyl phosphorothionate insecticides. Toxicol Apl Pharmacol 40:393-406, (1977b).

Matsumura, F., and Brown, A.W.A. Studies on carboxylesterase in malathion-resistant culex tarsalis. J Econ Entomol 56:381-388, (1963).

Matsunaka, S. Propanil hydrolysis:inhibition in rice plants by insecticides. Science 160:1360-1361, (1968).

Miles, J.W., Mount, D.L., Staiger, M.A., and Teeters, W. R. S-Methyl isomer content of stored malathion and fenitrothion water-dispersible powders and its relationship to toxicity. J Agric Food Chem 27:421-425, (1979).

Mirer, F.E., Levine, B.S., and Murphy, S.D. Parathion and methyl parathion toxicity and metabolism in piperonyl butoxide and diethyl maleate pretreated mice. Chem Biol Inter 17:99-112, (1977).

Murphy, S.D. Liver metabolism and toxicity of thiophosphate insecticides in mammalian, avian and piscine species. Proc Soc Exp Biol Med 123:392-403, (1966).

Murphy, S.D. Mechamisms of pesticide interactions in vertebrates. Residue Reviews, 25:201-221, (1969).

Murphy, S.D. Introductory remarks: Symposium on the role of biotransformation of non-hepatic microsomal mechanisms in altering toxicity. Toxicol Appl Pharmacol 23:738-740, (1972a).

Murphy, S.D. The toxicity of pesticides and their metabolites. In: Degradation of Synthetic Organic Molecules in the Biosphere. Procedings of a Conference of the National Academy of Sciences. NAS Press, Washington, D.C. pp313-335, (1972b).

Murphy, S.D. Assessment of the potential for toxic interactions among environmental pollutants. In: The Principles and Methods in Modern Toxicology, C.L. Galli, S.D. Murphy, and R. Paoletti, eds. Amsterdam: Elsevier/North-Holland Biomedical Press, 1980, pp 277-294.

Murphy, S.D. Toxic effects of pesticides. In: Toxicology-The Basic Science of Poisons, 3rd ed. C.D. Klaassen, M.O. Amdur, J.Doull, eds. Macmillan Pub Co., New York, pp519-581, (1986).

Murphy, S.D., Anderson, R.L., and DuBois, K.P. Potentiation of the toxicity of malathion by triorthotolyl phosphate. Proc Soc Exp Biol Med 100:483-487, (1959).

Murphy, S.D., Cheever, K.L., Chow, A.Y.K., and Brewster, M. Organophosphate insecticide potentiation by carboxylesterase inhibitors. Proc Europ Soc Tox XVII, Excerpta Medica Internat. Cong. 376:292-300, (1976).

Murphy, S.D., Costa, L.G., and Wang, C. Organophosphate insecticide interaction at primary receptors and secondary receptors. In: Cellular and Molecular Neurotoxicology, T. Narahashi. Raven Press, New York, pp165-176, (1984).

Murphy, S.D., and DuBois, K.P. Metabolic conversion of ethyl-p-nitrophenyl thionobenzene phosphonate (EPN) to an anticholinesterase agent. Fed. Proc. 15:462, (1956).

Murphy, S.D., and DuBois, K.P. Enzymatic conversions of the dimethoxy ester of benzotriazine dithiophospheric acid to an anticholinesterase agent. J. Pharmacol Exper Therap, 119:572-583, (1957).

Murphy, S.D., Lauwerys, R.R., and Cheever, K.L. Comparative anticholinesterase action of organophosphate insecticides in vertebrates. Toxicol Appl Pharmacol 12:22-35, (1968).

Myers, D.K., Mendel B., Gersmann, H.R., and Ketelaar, J.A.A. Oxidation of thiophosphate insecticides in the rat. Nature 170:805-807, (1952).

Nakatsugawa, T., and Dahm, P.A. Microsomal metabolism of parathion. Biochem Pharmacol 16:25-38, (1967).

Neal, R.A. Studies on the metabolism of diethyl 4-nitrophenyl phosphorothionate (parathion) in vitro. Biochem J 103:183-191, (1967).

Neal, R.A., and DuBois, K.P. Studies on the mechamism of detoxification of cholinergic phosphorothionates. J. Pharmacol Exp Ther 148:185-192, (1965).

Poore, R.E., and Neal, R.A. Evidence for extrahepatic metabolism of parathion. Toxicol Appl Pharmacol 23:759-768, (1972).

Shuphan, I., Sejull, Y., Rosen, J.D., and Casida, J.E. Toxicological significance of oxidation and rearrangment reactions of S-Chloroallyl thio and dithiocarbamate herbicides. In: Sulfur in Pesticide Action and Metabolism, ACS Symposium Series No.158, J.D. Rosen, P.S. Magee and J. E. Casida, eds. Washington, D.C.: American Chemical Society, 1981, pp65-82.

Singleton, S.D., and Murphy, S.D. Propanil (3,4-dichloropropionanilide)-induced methemoglobin formation in mice in relation to acylamidase activity. Toxicol Appl

Pharmacol 25:20-29, (1973).

Su, M., Kinoshita, F.K., Frawley, J.P., and DuBois, K.P. Comparative inhibition of aliesterases and cholinesterase in rats fed eighteen organophosphorus insecticides. Toxicol Appl Pharmacol 20:241-249, (1971).

Talcott, R.E., Mallipudi, N.M., Umetsu, N., and Fukuto, T.R. Inactivation of esterases by impurities isolated from technical malathion. Toxicol Appl Pharmacol 49:107-112, (1979).

Wang, C., and Murphy, S.D. Kinetic analysis of species difference in acetylcholinestease sensitivity to organophosphate insecticides. Toxicol Appl Pharmacol 66:409-419, (1982).

TOXICOLOGY OF ORGANOPHOSPHATES AND CARBAMATES

A. Moretto*
Albert Einstein College of Medicine
1300 Morris Park Avenue
Bronx, New York 10461, USA

M.K. Johnson
Toxicology Unit, MRC Laboratories
Woodmansterne Road
Carshalton, Surrey SM5 4 EF, ENGLAND

Toxicity of a chemical is dependant on several different but interconnected phases: a) factors which influence the delivery of the ultimate toxic agent to its site of action (absorption, distribution, storage, activation, detoxification); b) the reaction with the primary target (reversible or irreversible); c) the biochemical and physiological consequences and d) the clinical expression of toxicity. This paper will focus on the toxicity of organophosphorus (OP) ester insecticides in relation to the mechanism of their interaction with the primary targets, acetylcholinesterase (AChE) and Neuropathy Target Esterase (NTE). While OP-interaction with AChE causes acute effects (cholinergic syndrome), interaction with NTE leads to the development of a completely different syndrome known as OP-induced delayed polyneuropathy (OPIDP). The effect of carbamate insecticides on AChE will also be briefly discussed. Carbamates do not interact significantly with NTE and are structurally unable to cause OPIDP.

Many other effects, caused by single OP or by a group of related OPs, have been reported, but not all are well substantiated. These include behavioural and chronic effects on central nervous system, mutagenic, carcinogenic, teratogenic, and porphyric effects, effects on the immune system, on hormones, on the reproductive system, on the retina, on lipid metabolism, etc. These effects have been reviewed elsewhere (WHO, 1986a).

The toxic effects of non-anticholinesterase carbamate herbicides and fungicidal dithiocarbamates are reviewed elsewhere (WHO, 1986b) and will not be addressed here.

---

*present address: Istituto di Medicina del Lavoro
                  Universita' di Padova
                  via Facciolati 71
                  35127 PADOVA, ITALY.

NATO ASI Series, Vol. H13
Toxicology of Pesticides: Experimental, Clinical
and Regulatory Aspects. Edited by L. G. Costa et al.
© Springer-Verlag Berlin Heidelberg 1987

CHEMICAL PROPERTIES OF ORGANOPHOSPHORUS AND CARBAMATE INSECTICIDES

Table 1 illustrates various structures of OP insecticides. These chemicals are normally esters, amides, or thiol derivatives of phosphoric, phosphonic, phosphorothioic, or phosphonothioic acids. The general structural formula is:

$$R^1_{}\!\!>\!\!\underset{R^2}{\overset{O(S)}{\underset{\|}{P}}}\!\!-\!X$$

R^1 and R^2 are usually simple alkyl or aryl groups both of which may be bonded directly to the phosphorus atom (in phosphinates), or linked via -O-, or -S- (in phosphates), or R^1 may be bonded directly and R^2 via one of the above groups (phosphonates). In phosphoroamidates, carbon is linked to phosphorus through an -NH group. The group X can be any one of a wide variety of substituted and branched aliphatic, aromatic, or heterocyclic groups, linked to phosphorus via a bond of some lability (usually -O- or -S-), referred to as the leaving group. The double bonded atom may be oxygen or sulfur, and related compounds would be called phosphates or phosphorothioates (the nomenclature 'thiophosphate' or 'thionophosphate' is now outdated). The P=O form of a thioate ester, referred to as the oxon, is often incorporated in the trivial name (e.g., parathion is the parent compound of paraoxon).

Carbamate insecticides are usually esters of mono or dimethylcarbamic acid:

$$\underset{\text{H or CH}_3}{\overset{\text{H or CH}_3}{>}}\!\!N\!\!-\!\!\overset{O}{\overset{\|}{C}}\!\!-\!X$$

As for OPs, X may be one of a wide variety of groups linked to the carbonyl carbon atom via a labile ester or oxime-ester bond. Occasionally, long chain alkyl groups replace the methyl group. Complex labile groups may be attached to the alkyl group to form an inactive precursor which is metabolized to the proximal toxin in vivo.

ACUTE EFFECTS OF ORGANOPHOSPHORUS AND CARBAMATE INSECTICIDES: INHIBITION OF AChE

The acute effects of these insecticides are the inhibition of AChE in the nervous system and subsequent accumulation of toxic levels of endogenous acetylcholine (ACh) in nervous tissue and effector organs in both insects and mammals. In mammals, ACh is the chemical transmitter of nerve impulses at endings of post-

Table 1: Chemical structure of OP insecticide (modified from WHO, 1986a)

PHOSPHORUS GROUP AND COMMON NAMES	STRUCTURE
PHOSPHATE chlorfenvinphos, crotoxyphos, dichlorvos, dicrotophos, heptenphos, mevinphos, monocrotophos, naled, phosphamidon, TEPP, tetrachlorvinphos, triazophos	$(R\text{-}O)_2\text{-}\overset{\overset{O}{\|}}{P}\text{-}O\text{-}X$
S̲-ALKYL PHOSPHOROTHIOATE profenfos, trifenfos	$\underset{R\text{-}O}{\overset{R\text{-}S}{>}}\overset{\overset{O}{\|}}{P}\text{-}O\text{-}X$
S̲-ALKYL PHOSPHORODITHIOATE prothiofos, sulprofos	$\underset{R\text{-}O}{\overset{R\text{-}S}{>}}\overset{\overset{S}{\|}}{P}\text{-}O\text{-}XC$
O̲-ALKYL PHOSPHOROTHIOATE amiton, demeton-S-methyl, omethoate, oxydemeton-methyl, phoxim, vamidothion	$(R\text{-}O)_2\text{-}\overset{\overset{O}{\|}}{P}\text{-}S\text{-}X$
azothoate, bromophos, bromophos-ethyl, chlorpyrifos, chlorpyrifos-methyl, coumaphos, diazinon, dichlofenthion, fenchlorphos, fenitrothion, fenthion, iodofenphos, parathion, parathion-methyl, pyrazophos, pyrimiphos-ethyl, pyrimiphos-methyl, sulfotep, temephos, thionazin	$(R\text{-}O)_2\text{-}\overset{\overset{S}{\|}}{P}\text{-}O\text{-}X$
PHOSPHORODITHIOATE amidithion, azimphos-ethyl, azimphos-methyl, dimethoate, dioxathion, disulfoton, ethion, formothion, malathion, mecarbam, menazon, methidathion, morphothion, phenthoate, phorate, phosalone, phosmet, prothoate, thiometon	$(R\text{-}O)_2\text{-}\overset{\overset{S}{\|}}{P}\text{-}S\text{-}X$
PHOSPHOROAMIDATE cruformate, fenamiphos, fosthietan	$(R\text{-}O)_2\text{-}\overset{\overset{O}{\|}}{P}\text{-}NR_2$
PHOSPHOROTRIAMIDATE triamiphos	$\underset{R_2N}{\overset{R_2N}{>}}\overset{\overset{O}{\|}}{P}\text{-}NR_2$
PHOSPHOROTHIOAMIDATE methamidophos	$R\text{-}O\text{-}\underset{\underset{S\text{-alkyl}}{\|}}{\overset{\overset{O}{\|}}{P}}\text{-}NR_2$
isofenphos	$(R\text{-}O)_2\text{-}\overset{\overset{S}{\|}}{P}\text{-}NR_2$
PHOSPHONATE butonate, trichlorfon	$\underset{R}{\overset{R\text{-}O}{>}}\overset{\overset{O}{\|}}{P}\text{-}O\text{-}X$
PHOSPHONOTHIOATE EPN, trichlornat, leptophos, cyanofenphos	$\underset{R}{\overset{R\text{-}O}{>}}\overset{\overset{S}{\|}}{P}\text{-}O\text{-}X$

ganglionic parasympathetic nerve fibers, somatic motor nerves to skeletal muscle, pre-ganglionic fibers of both parasympathetic and sympathetic nerves, and certain synapses in the central nervous system. Thus, accumulation of ACh causes signs and symptoms that mimic nicotinic, muscarinic and central nervous system actions of ACh. The clinical picture, described in detail elsewhere (Plestina, 1984; Taylor, 1985; WHO 1986a,b) is summarized as follows:

a) muscarinic manifestations: increased bronchial secretion; excessive salivation, sweating and lacrimation; pinpoint pupils; bronchoconstriction; abdominal cramps; bradycardia.

b) nicotinic manifestations: fasciculation of fine muscles and, in more severe cases, of diaphragm and respiratory muscles; tachycardia.

c) central nervous system manifestations: headache; dizziness; restlessness; anxiety; mental confusion; convulsions; coma; depression of the respiratory centre.

These symptoms can occur in different combinations and time of onset, sequence, and duration may vary, depending on the chemical, dose, and route of exposure. Mild poisoning might include muscarinic and nicotinic signs only. Severe cases always show central nervous system involvement with a clinical picture dominated by respiratory failure, occasionally leading to pulmonary oedema.

Kinetic studies of inhibition of AChE by OP esters

The chemistry of OP inhibition of AChE and other esterases (e.g., NTE, liver carboxylesterase, serum butyrylcholinesterases, trypsin and chymotrypsin) is illustrated in Figure 1. Following the formation of a Michaelis complex (reaction 1), a specific serine residue in the protein is phosphorylated with loss of the leaving group X (reaction 2). Two further reactions are possible: reaction 3 (reactivation) may occur spontaneously at a rate that is dependent on the nature of the attached group and on the protein. It is also influenced by pH and added nucleophilic reagents, such as oximes, which may catalyse reactivation. Reaction 4 (aging) involves cleavage of an R-O-P bond with the loss of R and the formation of a charged monosubstituted phosphoric acid residue still attached to protein. The reaction is called aging because it is time-dependent, and the product is no longer responsive to nucleophilic reactivating agents. Each step is discussed in detail below. The interaction with carbamates is exactly analogous to reaction 1, 2 and 3 but reaction 4 is not possible. Also, while spontaneous reactivation (reaction 3) occurs with carbamates, oxime-catalysis of this reaction does not occur.

Inhibition

When cholinesterase is incubated with an OP or carbamate ester, the activity of the enzyme will decrease in a time-dependent manner and the rate of inhibition will be dependent on the inhibitor concentration (Figure 2). Addition of substrate can, however, halt or greatly reduce the rate of inhibition. Taken together, this indicates that inhibitors react progressively with the active centre of AChE (Aldridge and Reiner, 1972). Formation of inhibitor-enzyme complex with [inhibitor]≫[enzyme] is a pseudo-first-order bimolecular reaction described by

$$k_a = \frac{1}{[I]t} \ln (100/x)$$

where k_a = bimolecular constant rate; [I] = concentration of inhibitor; x = percentage of activity of the enzyme after treatment with inhibitor for various times; t = time of preincubation of enzyme and inhibitor before addition of substrate.

The rate constant k_a is a characteristic of the enzyme under defined condition of temperature, pH and composition of the medium and can be determined graphically as follows:

1) the above kinetic equation can be rearranged to: $\log x = \log 100 - k_a[I]t/2.303$
2) experiments are performed in which the quantity 'x' is determined after various times (t_1, t_2, etc) of incubation with a fixed concentration of inhibitor [I];
3) The values of 'log x' are plotted against 't' and should give a straight line with a slope = $-k_a[I]/2.303$, whence k_a may be calculated.

For the simplest case, the slopes of a family of such lines obtained using various concentrations of inhibitor should be proportional to [I] and all lines would intersect the ordinate at log 100% (= 2.0) (see Figure 2). Also, if a horizontal line is drawn at log 50% (1.7 on the graph), then the half-time ($t_{1/2}$) for inhibition at different concentrations can be read off the abscissa and should be inversely proportional to [I]. Departures from linearity or failure to intersect at 2.0 have been seen in a number of cases and provide valuable clues regarding the occurrence of spontaneous reactivation of inhibited enzyme (such as seen after inhibition by some OP esters and all simple carbamates) or other effects discussed in detail by Aldridge and Reiner (1972). It is also possible to obtain straight lines by plotting 'log x' vs [I] for fixed-time preincubations. The parameter [I_{50}] (that concentration which inhibits 50% of activity under defined conditions) can be read but, due to spontaneous reactivation, formation of reversible inhibitor-enzyme complexes, etc cannot be determined. It follows that I_{50} values obtained from experiments with only one preincubation time cannot be relied upon to indicate the true progressive inhibitory power of a compound: this procedure may lead either to under- or over-estimates of that power. Most of the compounds with biologically significant effects have a $t_{1/2}$ not higher than 20 min at 37°C for concentrations lower than 10^{-5} M (i.e., k_a=3,500 M^{-1} min^{-1}). Some compounds can be much more active, with a $t_{1/2}$=20 min at 10^{-9} M concentration.

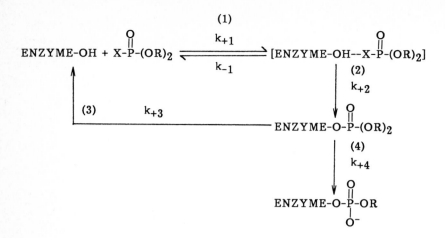

Fig. 1. Inhibition of an esterase by an OP compound. (1) Formation of Michaelis complex. (2) Phosphorylation of the enzyme. (3) Reactivation reaction. (4) "Aging".

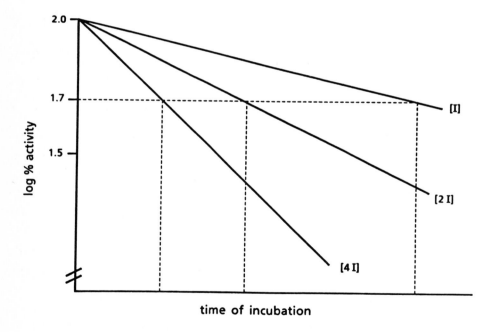

Figure 2: Decline of AChE activity following various periods of preincubation with an inhibitory OP or carbamate ester. The simplest case is illustrated but more complex sets of lines may be obtained (see text).

Reactivation

The classification of esters into substrates and inhibitors is somewhat arbitrary; the difference between them is in the velocity of reaction 3. Values of k_{+3} differ greatly between substrate and inhibitors. For the hydrolysis of ACh by AChE, k_{+3} is approximately 3×10^6 min^{-1} so that the acyl-enzyme is rapidly deacetylated and catalytic activity regenerated; for OP compounds and AChE, k_{+3} is 10^{-1}-10^{-6} min^{-1} and the regeneration of active enzyme is very slow. The rate of spontaneous reactivation of phosphorylated AChE depends on parameters such as pH and temperature, and also on the chemical structure of the side-chain bound to the phosphorus atom. For phosphorylated AChE, k_{+3} values vary according to R as follows: $(2\text{-ClEtO})_2 > (\text{MeO})_2 > (\text{IsoProO})_2 > (\text{EtO})_2$ (Reiner, 1971) and the half-life $(t_{1/2})$ of the inhibited enzyme ranges from less than 30 mins to more than 30 days. The presence of a sulfur instead of oxygen greatly increases the rate of reactivation of the analog compound (Clothier et al., 1981). It is believed that no spontaneous reactivation of AChE occurs after inhibition by phosphoroamidates, making such compounds intrinsically undesirable as insecticides. Reactivation of phosphorylated, but not carbamylated, AChE can be accelerated by nucleophilic agents such as NaOH, oximes $(R_2.C=NOH)$, hydroxiaminic acids $(R.CO.NHOH)$ or fluoride ions. Reactivation proceeds by the reaction of the nucleophile (water in the simplest case) with the electrophilic phosphorus attached to the enzyme. The reactivation of phosphorylated AChE by pralidoxime (2-PAM) is depicted in Figure 3. It should be noted that the phosphorylated oxime products of this reaction may be potent AChE inhibitors in themselves and very toxic, although they are usually very unstable.

The spontaneous reactivation of typical methylcarbamylated AChE of mammals proceeds faster than for most OP-inhibited AChEs. Half-lives of less than one hour are typical under physiological conditions.

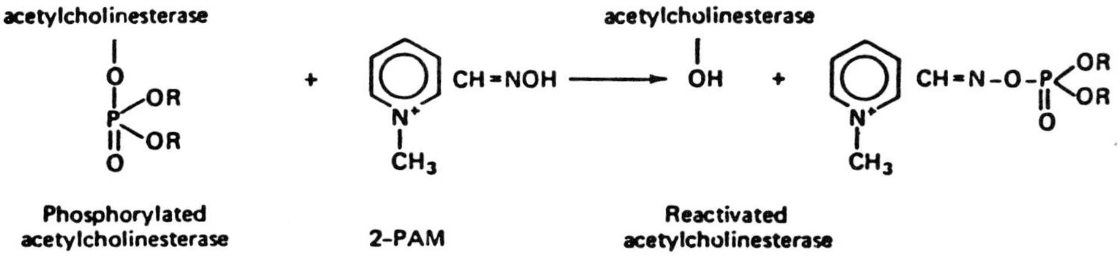

Figure 3: Reactivation of phosphorylated AChE by pralidoxime (2-PAM).

"Aging"

The aging phenomenon, which cannot occur with carbamates, is the time-dependent loss of ability of the phosphorylated enzyme to be reactivated by nucleophilic agents. The mechanism, described by the reaction 4, consists of cleavage of one R group and formation of a charged monosubstituted phosphoric acid residue on the protein. The rate of aging is, as for reactivation, dependent on pH, temperature etc, and also on the structure of the R group. For AChE the rate of aging is as follows: (highly branched alkyl group-O) >(Met-O) > (isoPro-O)>(Et-O) (O'Brien, 1967). When one or both of the residual alkyl groups are attached to the phosphorus through sulfur rather than oxygen, the rate of aging is increased, but not as markedly as the rate of spontaneous reactivation (Clothier et al., 1981).

In vivo extrapolation

Although in vitro data, described above, cannot be transposed directly to in vivo situations, they are consistent with the well known fact that, after poisoning by a sub-lethal dose of a carbamate or of some dimethyl phosphates, recovery and disappearance of symptoms is complete within a few hours. The value of k_{+3} for erythrocyte-AChE, taken from rats dosed in vivo with dimethyl phosphate, was reported to be 57×10^{-4} (a half life of inhibited enzyme of 2 hours) (Vandekar and Heath, 1957). One day after such a sub-lethal dose, most AChE was in the uninhibited form with a small fraction in the aged inhibited form. By contrast, no more than 20% of the inhibited enzyme would be expected to be reactivated one day after poisoning with diethyl phosphate (Vandekar and Heath, 1957) and a greater proportion of the inhibited enzyme would be aged, so that recovery to 100% activity would be very slow, depending mainly on the synthesis of fresh enzyme. It follows that markedly different outcomes would be expected after repeated intoxication with dimethyl phosphate and diethyl phosphate at doses capable of causing an initial 50% inhibition; the former would be less hazardous than the latter, because of a reduced possibility of 'build-up' of inhibited AChE. This concurs with the fact that rats can survive daily doses of 25% of the LD_{50} of trichlorfon (a dimethyl phosphate), but only about 12% of the LD_{50} of parathion (a diethyl phosphate) (DuBois, 1963).

Standard for a "safe" insecticide

The structure/activity relationships described above set standards which could be applied to the design of "safe" insecticides. Such standards would be as follows:

parameter	pest	man
(a) degradation and disposal	slow	fast

(b) interaction with AChE

(1) Affinity ($K_a = \dfrac{k_{-1} + k_{+2}}{k_{+1}}$)	low K_a	high K_a
(2) Conversion of Michaelis complex to acylated AChE (k_{+2})	high	low
(3) Instability of acylated AChE (k_{+3})	low	high
(4) Aging (k_{+4})	high	low

DELAYED NEUROPATHIC EFFECTS OF SOME ORGANOPHOSPHORUS INSECTICIDES

Regardless of the severity of anticholinesterase effects, some OPs can also induce a quite different syndrome known as OPIDP. Features of this syndrome are: (1) a delay of 1-3 weeks between ingestion of the agent and development of clinical signs: (2) the fact that a single dose may be sufficient to cause the effect even when the agent is unstable and cleared from the body within hours; (3) the preferential involvement of longer axons of spinal cord and peripheral nerves with degeneration in the most distal regions; (4) irreversible clinical effects in all but mildly-affected cases. Details of history, clinical and morphological features, and biochemical mechanisms are reviewed elsewhere (Johnson, 1975a; 1982; Bouldin and Cavanagh, 1979; Davis and Richardson, 1980; Lotti et al., 1984).

Organophosphorus esters and Neuropathy Target Esterase

The first essential step in the initiation of the delayed neuropathic effect of an OP is phosphorylation of a target protein in the nervous system. The protein, which was first identified by radiolabelling (Johnson, 1969), has esteratic activity and phosphorylation can be conveniently monitored as a progressive inhibition of the activity of this enzyme, now known as Neuropathy Target Esterase (NTE, formerly Neurotoxic Esterase) (Johnson, 1982). The second, and equally essential step, is aging of phosphorylated NTE. The steps of OP-NTE reactions are shown in Figure 1. As illustrated in Figure 4A, both inhibition **and** aging of NTE are essential for initiation of neuropathy. When NTE is inhibited by a suitable phosphate, phosphonate, or phosphoroamidate, aging is always possible, and occurs rapidly with a wide variety of neuropathic esters (Clothier and Johnson, 1980). However, after inhibition with phosphinates (which contain two phosphorus-carbon bonds) or sulphonyl fluorides, no hydrolysable bonds remain in the attached inhibitor molecule. Thus, aging is not possible (Figure 4B) and animals (hens) treated with these compounds do not develop OPIDP. Whenever NTE has been phosphinylated or sulphonylated in vivo, hens became resistant to challenge doses of typical neuropathic esters because the two-step initiation process had been blocked halfway. These animals are not resistant to anticholinesterase effects, if any, of challenge doses nor to other neuropathic agents such as acrylamide. Carbamates are either totally without effect (anticholinesterase carbamate are poor inhibitor of NTE) or inhibit NTE, but do not age. Thus, they protect the animal in the same way as the phosphinates. Phosphinates, sulphonyl fluorides, and carbamates exert their protective effect **if, and only if,** they are administered before the neuropathic agent (Johnson, 1970; 1974). Thus, it is clear that the consequences of OP-NTE interaction are quite different from those of AChE inhibition. In the latter case, acute toxicity arises directly from the loss of catalytic activity of AChE, leading to an accumulation of physiological substrate (ACh). On the other hand, initiation of OPIDP requires the generation of a certain quantity of modified (i.e. inhibited **and** aged) NTE in the nervous system at some point in time. Mere loss of NTE catalytic activity does not cause neuropathy; there is no evidence of deleterious accumulation of physiological substrate or lack of hydrolysis products, as demonstrated by the lack of neuropathic effect after NTE inhibition by phosphinates, sulphonyl fluorides or carbamates. However, the loss of catalytic activity is conveniently used as an indicator of covalent binding to NTE, but initiation of OPIDP is possible only if inhibition is followed by aging (i.e., the biological response depends on the nature of the group covalently bound to the esterase active site). In adult hens, detectable neuropathic events are never seen after a single dose/exposure of an OP unless at least 70% of the normally available NTE is converted to the charged modified form. In single-dose experiments, the peak amount (measured as loss of NTE catalytic activity) is usually reached within 1-72 hours of dosing: the time depends on the speed of metabolic activation (if any) and disposal reactions for the particular compound. Owing to the synthesis of fresh protein, this inhibition declines markedly during the 8 to 14-day delay period, and there is no correlation between neuropathy and NTE inhibition measured at the time clinical signs reach their peak. The sequence of events leading from formation of this quantum of modified NTE to axonal degeneration and clinical neuropathy is largely unknown (see Moretto et al., in this book).

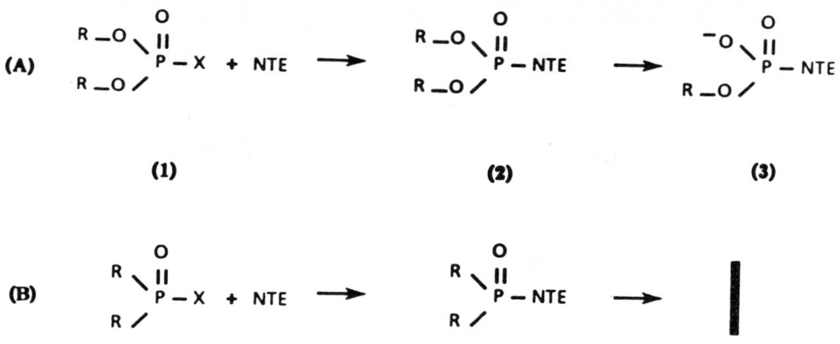

Figure 4: Reaction of NTE with an organophosphate (A) or an organophosphinate (B). **1.** enzymatic activity, no known physiological role; **2.** inhibited enzyme, no toxic effect; **3A.** aged inhibited enzyme, initiation of the mechanism of OPIDP; **3B.** no aging possible; initiation of the mechanism of OPIDP impossible, blockade of the effect of neuropathic OPs.

Prediction of neuropathic effects and extrapolation to man

In vitro studies of NTE and in vivo tests as listed by Johnson (1975b) reveal a number of factors which affect the neuropathic potential of a compound. Structure/activity relationships were summarized by Johnson (1982) as follows:

(A) Factors that increase delayed neurotoxicity potential more than acute toxicity are:

(1) choice of phosphonates or phosphoroamidates rather than analogous phosphates;

(2) increase in chain-length or hydrophobicity of R^1 and R^2;

(3) a leaving group X, which does not sterically hinder the approach to the active site of NTE.

(B) Factors that decrease the comparative potential are:

(1) the converse of (A) 1, 2, and 3;

(2) the choice of R or X groups that are very bulky (naphthyloxy) or non-planar;

(3) the choice of a nitrophenyl group at X (a steric effect?);

(4) choice of comparatively more hydrophilic X groups (oximes or heterocyclics);

(5) choice of thioether linkages at X.

Based on the structure/activity relationships for NTE described above together with those previously described for AChE, it can be then understood why a) malathion (O,O-dimethyl-S-(1,2-dicarbethoxyethyl) phosphorodithioate) and diazinon (O,O-diethyl-2-isopropyl-6-methyl-4-pyrimidinyl phosphorothioate) are not able to cause

OPIDP, b) in its homologous series, only dichlorvos (O,O-dimethyl-2,2-dichlorovinyl phosphate) is not neuropathic at the LD_{50}, c) EPN (a phosphonothioate with a hydrophobic phenyl group at R^1) is neuropathic, even with a 4-nitrophenyl leaving group, and d) other phenylphosphonothioates such as desbromoleptophos, or cyanofenphos are also neuropathic. Lotti and Johnson (1978) suggested another way to predict the neuropathic potential of OPs utilizing in vitro studies with the target enzymes. Both AChE and NTE are, in fact, associated with nervous tissue and it seems reasonable to propose that whatever percentage of a dose which ultimately reaches the nervous system in the active form (after absorption, metabolic activation and/or deactivation, distribution and/or excretion etc.), would prefer to react with AChE or NTE according to the relative potencies demonstrable with these enzymes in vitro. Their experiments showed that the comparative inhibitory power of OPs in vitro against hen brain AChE and NTE correlates with their comparative effects seen (death or OPIDP) in vivo. The characteristics of NTE in hens and humans are very similar (Lotti and Johnson 1978; 1980a) as is true also for AChE, so extrapolation from in vitro to in vivo ratios in humans seems justifiable to predict which toxic effects are likely to predominate after exposure to a given OP. Such extrapolations assume that the same amount of inhibition of NTE is necessary to trigger the mechanism in humans as in the experimental animal. What is not known, however, is the numerical value of the threshold of inhibition of NTE in man that is associated with clinical neuropathy. Individuals poisoned with trichlorfon (dichlorvos being the active metabolite) and treated only with atropine, survived and developed neuropathy (Johnson, 1981). By contrast, trichlorfon neuropathy can be produced in hens only after huge doses coupled with prophylaxis and therapy for severe anticholinesterase effects (Johnson, 1981). It seems, therefore, in this case, a severely poisoned man has a higher chance of developing neuropathy than of dying, compared to the hen. It might be concluded that human NTE is more sensitive to in vivo inhibition by this compound compared to AChE, as suggested by data from in vitro studies (Lotti and Johnson, 1978). Another possibility is that the threshold level of NTE inhibition for initiation of OPIDP in man is less than the 70% value for hen, but there is no way of directly verifying this.

DIAGNOSIS OF HUMAN POISONING (see also WHO 1986 a,b)

Based on data obtained from studies on the biochemical action of OPs, a relatively specific clinical test was designed for diagnosis of excessive exposure to these compounds. AChE is present in human erythrocytes (RBC) and is very similar to the enzyme present in target synapses. Thus, levels of RBC-AChE are assumed to mirror the effects in target organs, but only when the OP has equal access to blood and synapses. In the case of acute poisoning, a high inhibition of RBC-AChE is pathognomonic, but in the follow-up of the intoxication, it might not be correlated with the severity of symptoms. The recovery of AChE activity is due to the combined effects of reactivation (spontaneous or induced) and synthesis of new enzyme: also, the life of RBCs is 120 days so that synthesis of new RBC-AChE is only due to new RBC production.

Determination of AChE inhibition due to carbamate intoxication in vivo poses some technical difficulties. It is possible to observe typical cholinergic symptoms following carbamates but, unless precautions are taken, by the time tissues are removed and prepared for assay, reactivation of carbamylated AChE may have occurred and inhibition is undetectable.

Blood-plasma contains an enzyme related to AChE, usually called pseudo-ChE or butyrylcholinesterase. It has no known physiological function and can be inhibited selectively by some compounds without causing a toxic response; the sensitivities of AChE and pseudo-ChE to inhibitors differ, so that measurement of total blood or plasma ChE activity is to be considered less reflective of nerve tissue activity (Hayes, 1982).

NTE activity has been found in blood lymphocytes of both hen and man, and the possibility of monitoring exposed individuals by means of lymphocytic NTE measurements is being explored (Lotti et al., 1983; Bertoncin et al., 1985; Lotti et al., 1986; Maroni and Bleeker, 1986). In one case the development of OPIDP was predicted two weeks before clinical and electrophysiological tests indicated that nerve degeneration had occurred. The prediction was based on reduced levels of lymphocytic NTE activity found in the intoxicated man after signs of cholinergic poisoning developed (Lotti et al, 1986).

RATIONALE FOR TREATMENT OF HUMAN POISONING

The understanding of the mechanism of acute toxicity of OP insecticides has provided a basis for rational therapy of OP poisoning. The effects of inhibition of AChE, as previously described, are common to all OP and carbamate intoxication. However, the speed of onset and the rate of unaided recovery from sub-lethal doses vary greatly, depending on the chemical nature of the pesticide, the route of exposure, and on the characteristics of exposure (one ovelwhelming dose or repeated small doses over time).

Factors leading to a slow onset of symptoms include:

a) slow absorption or metabolism; and

b) persistence in the system of a comparatively stable inhibitor of AChE (either as low concentrations of an active inhibitor or as high concentrations of a weak inhibitor).

Factors leading to rapid clearance of symptoms include:

a) rapid clearance of the pesticide and its active metabolites;

b) a slow rate of aging of inhibited AChE which gives opportunity for reactivation (spontaneous or induced) to occur; and

c) rapid spontaneous reactivation of inhibited AChE.

Therapy of OP poisoning involves the use of an anticholinergic drug (usually atropine which antagonizes muscarinic and some central effects) plus a reactivator of inhibited AChE (an oxime) along with other supportive treatments and artificial respiration, if necessary. Diazepam, but not anticonvulsants, is a valuable adjunct to atropine/oxime in severe anticholinesterase intoxication and the tranquilizing effect may be useful in less severe cases. While atropine is effective as long as cholinergic signs are present, the use of reactivators of inhibited AChE is effective only on non-aged inhibited enzyme. Thus, oxime therapy becomes less effective with time after poisoning according to the rate of aging of inhibited AChE. It is therefore important to know which compound is involved and its behaviour with respect to inhibition, reactivation, and aging of AChE as described previously. The efficacy of oxime therapy is obviously prolonged in the case of compounds which are slowly absorbed, activated or released from storage tissues. It is emphasized that the benefit of joint use of atropine and oxime may far exceed the effect of either of the agents alone.

For carbamate poisoning, the use of oxime is contraindicated, probably because of the rapid reversal of the carbamate inhibited enzyme.

No specific therapy has yet been found against OPIDP.

CONCLUSIONS

The knowledge of the mechanism of the acute toxicity of OPs and carbamates has provided a rationale for resolving problems of human safety during the production and use of these insecticides. The structure/activity relationships described herein set guidelines for the design and production of "safe" insecticides. Understanding of the events following AChE inhibition by the active compound led to the development of specific clinical tests for the diagnosis of excessive exposure and to the understanding of the potentials as well as the limitations of these tests. A rational and specific therapy of poisoning was also developed. The molecular target of OPs which cause delayed neuropathy has been identified and progress has been made in understanding of the mechanism of initiation. NTE assays are beneficial in the evaluation of new compounds: every dose tested has a quantifiable response with regard to the threshold effect on NTE (70-80% inhibition and aging) required to initiate neuropathy; the response to chronic low-level exposure can be assessed (Lotti and Johnson, 1980b); and NTE studies in accessible tissues may be useful in health monitoring of exposed humans.

REFERENCES

Aldridge WN, Reiner E (1972) Enzyme inhibitors as substrates. North-Holland, Amsterdam/London.

Bertoncin D, Russolo A, Caroldi S, Lotti M (1985) Neuropathy target esterase in human lymphocytes. Arch Environ Health 40: 139-144.

Bouldin TW and Cavanagh JB (1979) Organophosphorus neuropathy. I. A teased-fiber study of the spatio-temporal spread of axonal degeneration. Am J Path 94: 241-252.

Clothier B, Johnson MK (1980) Reactivation and aging of neurotoxic esterase inhibited by a variety of organophosphorus esters. Biochem J 185: 739-747.

Clothier B, Johnson MK, Reiner E (1981) Interaction of some trialkyl phosphorothiolates with acetylcholinesterase: characterization of inhibition, aging, and reactivation. Biochem Biophys Acta 660:306-316.

Davis CS, Richardson RJ (1980) Organophosphorus compounds. In: Spencer PS, Schaumburg HH (eds) Experimental and Clinical Neurotoxicology. Williams and Wilkins, Baltimore, pp. 527-544.

DuBois KP (1963) Toxicological evaluation of the anticholinesterase agents. In: Koelle GB (ed) Handbook of Experimental Pharmacology. Springer-Verlag, Berlin, vol. 15, pp. 833-857.

Hayes JH Jr (1982) Pesticides Studied in Man. William and Wilkins, Baltimore/London, pp. 284-435.

Johnson MK (1969) The delayed neurotoxic effect of some organophosphorus compounds: identification of the phosphorylation site as an esterase. Biochem J 114: 711-717.

Johnson MK (1970) Organophosphorus and other inhibitors of the brain "neurotoxic esterase" and the development of delayed neurotoxicity in hens. Biochem J 120: 523-531.

Johnson MK (1974) The primary biochemical lesion leading to the delayed neurotoxic effects of some organophosphorus esters. J Neurochem 23: 785-789.

Johnson MK (1975a) The delayed neuropathy caused by some organophosphorus esters: mechanism and challenge. Crit Rev Toxicol 3: 289-316.

Johnson MK (1975b) Organophosphorus esters causing delayed neurotoxic effects: mechanism of action and structure/activity studies. Arch Toxicol 34: 259-288.

Johnson MK (1981) Do trichlorphon and/or dichlorvos cause delayed neuropathy in man or in test animals? Acta Pharmacol Toxicol Scand Suppl 5: 87-98.

Johnson MK (1982) The target for initiation of delayed neurotoxicity of organophosphorus esters: biochemical studies and toxicological applications. In: Hogson B, Bend JR, Phillip RM (eds) Reviews in Biochemistry and Toxicology. Elsevier, Amsterdam/Oxford/New York vol. 4, pp. 141-212.

Lotti M, Johnson MK (1978) Neurotoxicity of organophosphorus pesticides: predictions can be based on in vitro studies with hen and human enzymes. Arch Toxicol 41: 215-221.

Lotti M, Johnson MK (1980a) Neurotoxic esterase in human nervous tissue. J Neurochem 34: 747-749.

Lotti M and Johnson MK (1980b) Repeated small doses of a neurotoxic organophosphate. Monitoring of neurotoxic esterase in brain and spinal cord. Arch

Toxicol 45: 263-271.

Lotti M, Becker CE, Aminoff MJ, Woodrow JE, Seiber JN, Talcott RE, Richardson RJ (1983) Occupational exposure to the cotton defoliants DEF and Merphos. J Occup Med 25: 517-522.

Lotti M, Becker CE, Aminoff MJ (1984) Organophosphate induced delayed polyneuropathy: pathogenesis and prevention. Neurology 34: 658-662.

Lotti M, Moretto A, Zoppellari R, Dainese R, Rizzuto N, Barusco G (1986) Inhibition of lymphocytic Neuropathy Target Esterase predicts the development of Organophosphate induced delayed polyneuropathy. Arch Toxicol 59: 176-179.

Maroni M, Bleecker ML (1986) Neuropathy target esterase in human lymphocytes and platelets. J Appl Toxicol 6: 1-7.

O'Brien RC (1967) Insecticides: Action and Metabolism. Academic Press, New York, pp. 32-54.

Plestina R (1984) Prevention, diagnosis, and treatment of insecticide poisoning. World Health Organization, Geneva (report no. VBC/84.889).

Reiner E (1971) Spontaneuos reactivation of phosphorylated and carbamylated cholinesterase. Bull WHO 44: 109-112.

Taylor P (1985) Anticholinesterase agents. In Goodman LS, Gilman A (eds) The Pharmacological Basis of Therapeutics, 7th ed, Macmillan, New York, pp. 100-123

Vandekar M and Heath DF (1957) The reactivation of cholinesterase after inhibition in vivo by some dimethyl phosphate esters. Biochem J 67: 202-208.

WHO (1986a) Organophosphorus Insecticides: A General Introduction. World Health Organization, Geneva (Environmental Health Criteria 63).

WHO (1986b) Carbamates Insecticides: A General Introduction. World Health Organization, Geneva (Environmental Health Criteria 64).

TOXICOLOGY OF CHLOROPHENOXY HERBICIDES AND THEIR CONTAMINANTS

S. D. Murphy, Ph.D.

Department of Environmental Health, SC-34

School of Public Health and Community Medicine,

University of Washington

Seattle, WA 98195

The compounds 2,4-dicholorophenoxy acetic acid (2,4-D) and 2,4,5-trichlorophenoxy acetic acid (2,4,5-T) and their salts are among the most commonly recognized herbicides. These compounds and their salts and esters have long been used for control of broad leaf weeds and woody plants in which they exert a herbicidal action by acting on plant growth hormones. It is not known that they exert any hormonal action in animals, but their mechanisms of mammalian toxicity is still poorly understood. These compounds' toxicities have been comprehensively reviewed. (IARC, 1977; Hayes, 1982; IPCS, 1984; Crump, 1986).

The acute toxicities of the chlorophenoxy herbicides and their various esters and salts range from LD_{50}'s of 300 to 2000 mg per kilogram in several experiemental species that have been tested. There are reports that dogs are relatively more sensitive to the 2,4,5-T isopropyl ester, with an LD_{50} of around 100 mg per kilogram. With less than fatal doses, adverse effects observed in laboratory animals included anorexia, weight loss, myotonia, and various pathological changes in the gastrointestinal tract, lungs and liver. The toxic signs reported for the chlorophenoxy acetic acid herbicides are qualitatively similar for all compounds. Where as the no effect level (NOEL) for a single oral dose of 2,4-D in dogs was 25 mg/kg, monkeys tolerated 214 mg/kg by the oral route.

In general, adverse effects in animals exposed to acutely toxic doses of the chlorophenoxy herbicides include myotonia, anorexia, excessive thirst, depression, roughness of coat, tremors, myasthenia, ataxia, rapid breathing, salivation, paralysis and coma, with death resulting from ventricular fibrillation. Autopsy findings include stomach irritation, liver and kidney damage, and occasional lung congestion. Similar signs and symptoms have been reported in humans who have had heavy occupational exposure or accidental or suicidal ingestion of chlorophenoxy herbicides. it has been estimated that the single oral dose required to produce symptoms in humans is probably around three to four grams (40-60 mg/kg). Neuromuscular signs and symptoms have been reported in human exposures as well as in laboratory animal experiments. These include: profound muscular weakness, myotonia and

NATO ASI Series, Vol. H13
Toxicology of Pesticides: Experimental, Clinical
and Regulatory Aspects. Edited by L. G. Costa et al.
© Springer-Verlag Berlin Heidelberg 1987

peripheral neuritis. Feeding studies in laboratory animals indicate that repeated exposure to doses that are just slightly smaller than the single toxic doses are tolerated with respect to the usual kinds of chemical toxicity.

There have been numerous fatal or severe human poisonings by the chlorophenoxy herbicides. Signs and symptoms observed in these cases indicate involvement of several organ systems: the central nervous system, peripheral nervous system, skeletal muscle, digestive system, respiratory system, circulatory system, liver, kidney, skin and reproductive system.

In reports of occupational injury or illness associated with agricultural chemicals, dermatological conditions are the most frequently cited adverse effects among workers exposed to herbicides. The chlorophenoxy herbicides produce contact dermatitis in humans, a rather severe type of dermatitis known as chloracne has been observed in workmen involved in the manufacture of 2,4,5-T (Poland, et al., 1971). This effect appears due primarily to the action of a contaminant, 2,3,7,8-tetrachlorodibenzo-p-dioxin. During the past 15 years, concern about the toxicology of the chlorophenoxy herbicides, particularly 2,4,5-T, has been one of controversy with respect to whether effects are due to the herbicides themselves or to the polychlorinated dibenzodioxin contaminants.

METABOLISM AND DISPOSITION

Studies of the metabolism and distribution of chlorophenoxy herbicides indicate that absorption from the gastrointestinal tract occurs only with the free acids. When esters of these compounds were administered, no esters were observed in the blood or urine, and only free acids or their conjugated metabolites appeared. Studies of the rates of elimination of the chlorophenoxy herbicides following administration of single or a few doses to both laboratory animals and to humans indicate a fairly rapid elimination. Humans are intermediate between rats and dogs with regard to the rate of elimination of 2,4,5-T. The slower rate of elimination of 2,4,5-T in dogs may at least partially explain the greater sensitivity of this species to the acute toxicity of these compounds. Pharmacokinetic studies in human volunteers have indicated that the biological half-lives of 2,4,-D and 2,4-T are on the order of one to three days (Piper et al., 1973; Gehring, et al., 1973).

CONTAMINANT TOXICITY

As indicated earlier, the toxicology of the chlorophenoxy herbicides is complicated by the presence, or possible presence, of polychlorinated dibenzodioxins (PCDD) contaminants. The 2,3,7,8 isomer of tetrachloro dibenzo p-dioxin, or TCDD, is one of the most acutely toxic man-made chemicals known, with LD_{50}'s in the most sensitive mammalian species on the order of a few micrograms per kilogram. However, there is a fairly wide variation in species sensitivity to TCDD, with guinea pigs being the most susceptible (oral LD_{50} 0.6 microgram/kilogram) and hamsters the most resistant (1,157 micrograms/kilogram). The nature of the effects of TCDD in laboratory animals and humans has been reviewed by IARC (1977) and by Homberger et al., (1979). The toxic effects of repeated doses in laboratory animals include: systemic lesions involving liver, blood and skin, reduced fertility, fetotoxicity and teratogenicity, immunodeficiency, liver enzyme induction and carcinogenicity. The lethal action of TCDD is delayed, and deaths are seldom observed before 10 days to two weeks following a single dose. The nature of the signs are rather nonspecific with a type of wasting disease and loss of body fat becoming apparent after the first five to six days. The mechanism of the fatal acute toxicity of TCDD remains unknown. Reduced fertility in laboratory animals is among the most sensitive tests for an effect of TCDD. In repeated exposure studies altered reproductive capacity at daily dosages as low as 10 nanograms per kilogram has been reported. Immunodeficiency is also a sensitive indicator of TCDD effects in laboratory animals, occurring with about 100 nanograms per kilogram doseage. Although several experimental studies have now established that exposure to TCDD induces an increased incidence of several types of tumors in laboratory animals (at less than microgram per kilogram dosages), the evidence of mutagenic potential for TCDD is inconsistent and controversial.

With regard to humans, the most consistent finding among persons exposed to TCDD is the dermatologic condition known as chloracne. This is apparently also the most sensitive indicator and is often taken as specific evidence of exposure to TCDD or closely related compounds. Other conditions with dermatologic involvement that have been associated with human exposures to TCDD include porphyria cutanea tarda, hyperpigmentation, and hirsutism, all resulting from systemic exposures.

Industrial exposures to TCDD have occurred in factories manufacturing 2,4,5-T or other materials in which trichlorophenol is a precursor. Chloracne has frequently been associated with these exposures. In some cases, this condition persisted for as long as 20 years after the initial accidental exposure. Other signs of systemic poisoning detected by clinical chemistry tests suggest the liver as a target organ. This is consistent with observations in acute and subchronic toxicity studies in animals. Polyneuropathies were also reported for several of the individuals exposed in these industrial accidents. In a case of three scientists that were poisoned in the course of experimental preparation of TCDD, personality changes, loss of energy, impairment of vision, taste and muscular coordination and sleep disturbances, were delayed symptoms that were felt to be associated with TCDD exposure which had occurred approximately two years previously. The signs and symptoms reported for workers exposed to TCDD in industrial or other kinds of accidents, may have other causes, including the possibility of exposure to other chemicals. However, the chloracne, the liver injury, and the neurologic changes are consistent with observations in laboratory animals, and with observations made among workers exposed to 2,4,5-T which may have been contaminated extensively with TCDD.

NEUROLOGIC EFFECTS OF CHLOROPHENOXY COMPOUNDS

Peripheral neuropathy has been reported in several epidemiologic studies that have been conducted among workers who have had opportunity for exposure to 2,4,5-T, trichlorophenol or related polychlorinated dioxins. For example, Suskind (1953) conducted a follow-up study of affected workers that had been exposed in a 2,4,5-T production area. Thirty-five workers had shown chloracne and in the follow-up studies symptomes of peripheral neuropathy were found in 27 cases. Other neurologic signs included nervousness, fatigue, irritability, and decreased libido. Bauer et al., (1961) found fatique, muscle weakness, and pain mostly in the lower extremities as prevalent symptoms in workers who had developed chloracne while employed in the manufacture of 2,4,5-T. Sensory neuropathy as well as decreased ability to concentrate, memory deficits, decreased drive, and sleep disturbances were also observed. Forty-two workers exposed in an accidental venting of steam during the manufacture of trichlorophenol were studied by Goldman (1972). Seven workers showed signs or symptoms of central nervous system involvement and three exhibited polyneuropathy. Pazderova-Vijlupkova et al., (1981) reported a 31% incidence of abnormal nerve conduction velocities in 55 of 80 workers that were employed at a 2,4,5-T manufacturing plant. In one severely affected worker, some Schwann cell pathology was found at autopsy. Central nervous system symptoms were also observed, and thirty-six of the workers who were examined by a psychiatrist were characterized as "neurasthenic" or "depressive."

There are also a few reports of neurologic disorders among workers that have been exposed to 2,4-D. Goldstein et al., (1959) observed three cases of severe sensory and motor disorders in individuals exposed dermally to 2,4-D used as a spray. Cases of neuropathy following exposure to formulated 2,4-D products have also been reported by several other investigators (Todd, 1962; Berkeley and Magee, 1963; Berwick, 1970; and Wallis et al., 1970).

Singer et al., (1982) studied nerve conduction velocities of fifty-six workers with a mean age of thirty-five years and a 7-year mean duration of employment in the manufacturing of chlorophenoxy herbicides. Findings were compared with a control group of twenty-five subjects without exposure to neurotoxic agents. Significantly slowed conduction was noted in the sural and the median motor nerves of exposed workers as compared with conduction veolocities in controls. The duration of employment was correlated with the degree of slowing of the sural nerve conduction velocity. Forty-six percent of the individuals in the exposure group had one or more slowed nerve conduction velocities as compared to only 5% of the selected control group.

There has been relatively little attention paid to possible peripheral neuropathic effects in laboratory animals associated with exposure to chlorophenoxy herbicides or the dioxins. Descriptive reports of the toxic affects 2,4-D indicate that motor disorders, paralysis of the extremeties, and myatonia are observed in a variety of laboratory species (Bucher, 1946; Hill and Carlisle, 1947; and Desi et al., 1962). Electroencephalographic abnormalties as well as histopathologic evidence of nervous system susceptibility to 2,4,-D has also been reported by Elo and Ylitalo (1979) and Way (1969). Blakely and Schiefer (1986) reported that acute and subacute dermal dosing of mice caused perivascular edema and ganglion cell necrosis in the central nervous system.

REPRODUCTIVE TOXICITY AND TERATOGENIC EFFECTS

During the 1970's, concern about the toxicology of 2,4,5-T and related compounds centered primarily on the apparent teratogenic action in experimental animals (Courtney and Moore, 1971). It was subsequently determined that those studies, which indicated fetotoxic and teratogenic effects of 2,4,5-T, were conducted with a sample of the herbicide that contained a high level (about 30 parts per million) of a contaminant, namely the 2,3,7,8 TCDD (Panel on the Herbicides, 1971). Because of the extremely high acute toxicity of TCDD, it is apparent why reported effects of 2,4,5-T observed in both laboratory and field experience might be attributable to the dioxin contaminant. For example, for female guinea pigs, the ratio of the LD_{50} of 2,4,5-T to the LD_{50} of TCDD is 630,000. For female rats, the acute oral LD_{50} for TCDD given to pregnant rats during the gestational period that resulted in fetal toxicity was only about 1/400 of the maternal of LD_{50} dioxin, or it can be calculated to be one four-millionth of the single oral LD_{50} of 2,4,5-T to female rats (Murphy, 1972). Clearly then, the concentration of TCDD contaminant in a technical sample of 2,4,5-T could be a major factor in determining its fetotoxicity or teratogenicity. These early experiments on the teratogenesis and fetotoxicity of 2,4,5-T illustrated an important principle in the process of evaluating the safety of commercial products; namely, that one must be concerned not only with the major active component, but also with minor contaminants that may be present due to formation during manufacture or from degradative reactions occurring in the environment.

Subsequent to the finding of an extremely potent teratogenic and fetotoxic contaminant in 2,4,5-T, numerous experiments have been conducted to test for a possible teratogenic action of relatively uncontaminated samples (generally containing less than 0.5 ppm TCDD). These experiments showed that with sufficient doses, that is of the order of 50 to 100 mg per kilogram per day during organogenesis, purified 2,4,5-T did produce cleft palate and cystic kidney and some fetotoxic effects in mice and hamsters. However, rats and monkeys appeared resistant to the teratogenicity of 2,4,5-T itself (Hayes, 1982).

Various studies of the teratogenicity and fetotoxicity of 2,4-D have yielded inconsistent results and generally this herbicide is less suspect of teratogenic action than the trichlorinated analog (IPCS, 1984). The potent fetotoxic agent 2,3,7,8 TCDD is not a contaminant of 2,4-D, although some related, less toxic, dioxins may be present in low concentrations. Of the positive findings of teratogenic action of 2,4-D in experimental studies, the abnormalities generally consisted of increased incidence of fused ribs, distorted scapula, defects in bones of fore and hind limbs, and micromelia (Khera and McKinley, 1972; Schwetz et al., 1971). Using the data from some of the reports of teratogenic effects of 2,4-D reported in the literature, a NOEL, (no observed effect level) for reproductive effects has been estimated by Crump et al., (1986) to be 25 mg/kg/day for rats. Considering the estimated worse case human exposures resulting from a single spraying episode, they estimated that with this NOEL the margin of safety for teratogenic effects is about 150 for the most heavily exposed occupational groups and from 500 to 78,000 for environmental exposures to different media that might be contaminated with 2,4-D. These estimations of margins of safety from worst case conditions suggest that the hazard from 2,4-D with respect to tgeratogenic effects is rather low.

There has been considerable public concern about the possibility of teratogenic effects resulting from environmental exposure to chlorophenoxy herbicides or related chlorinated dioxin contaminants. Epidemiologic studies designed to evaluate the risk of teratogenic affects of these herbicides and their contaminants have failed to provide unequivocal evidence that there is increased teratogenic risk involved with normal use of the herbicide. Exaggerated conditions of potential exposures as might have occurred with the use of agent orange in Vietnam or the community contamination with TCDD from the Icmesa chemical plant accident in Seveso, Italy have stimulated investigations. Homberger et al., (1979), indicated that in the Seveso area, there was no greater incidence of spontaneous abortions, among residents of the contaminated area, in any one of the four quarters of the year following the exposures, when compared to the incidence in areas outside the contaminated zones. An epidemiological investigation reported by Hamify et al., (1981) found no association of spraying 2,4,5-D in New Zealand with the incidence of spina bifida, cleft lip, or cleft palate (the usual malformations seen in rodent studies). However, there was a statistical association with exposure and congenital abnormalities of the foot and the urethral opening.

The Center for Disease Control in Atlanta, GA (Erickson, et al., 1984) recently completed an extensive case control study to determine whether cases of malformations occurred with higher frequency among babies whose fathers were Vietnam veterans, as compared to other veterans who did not have opportunity for exposure to agent orange. The study examined four questions: whether veterans (excluding Vietnam veterans) were at different risk for fathering babies with defects than non-veterans; whether Vietnam veterans were at different risk for fathering babies with defects (the central question); whether Vietnam veterans who received higher Agent Orange Exposure Opportunity Index scores were at different risk than other men; and whether Vietnam veterans who said they believed they had been exposed to agent orange were at different risk than other men. Each of these hypotheses were evaluated for ninety-six different defect groups. There were 428 case-group babies and 268 control-group babies considered for Vietnam veterans and 4,387 case-group babies and 2,699 control-group considered for non-Vietnam veterans.

When all incidents of malformation were considered, the odds ratio was 0.97 with 95% confidence limits of 0.83 to 1.14. This provided no support to the hypothesis that Vietnam veterans are at increased risk, in general, for fathering deformed babies than other men. However, the estimated risk for fathering babies with the specific malformation of spina bifida was higher for Vietnam veterans which had a higher Agent Orange Opportunity Index Score. These veterans also had higher estimated risk for fathering babies with cleft lip and with the broadly categorized defect "other neoplasms." Although the investigators (and others) who have reviewed these studies do not interpret these findings as strongly supportive of a risk of malformed children in agent orange-exposed veterans, the few circumstances which do seem to be associated with higher opportunity for agent orange exposure will likely continue to plague both concerned veterans and public health personnel. However, a similar study of Australian veterans who served in Vietnam led to the same conclusion of "no evidence that Army service in Vietnam relates to the risk of fathering a child with an anomaly " (Donovan et al., 1983). It is worth nothing in this regard that animal studies in which male rodents were exposed individually and in combinations to 2,4-D, 2,4,5-T and the contaminant TCDD and mated with unexposed females, also failed to reveal effects on the offspring (Lamb et al., 1981).

IMMUNOTOXICITY

The herbicide 2,4,5-T is among ten pesticides that Vos et al., (1983) concluded had a "marked" effect on the immune system. Immunosuppressant effects also have been observed to occur in laboratory animals exposed to doses of TCDD below those responsible for most other chemical toxicity effects (Vos et al.,1974).

In a recently reported epidemiologic study by Hoffman et al., (1986) residents of a mobile home park, in which the roadside soil had been heavily contaminated with TCDD, were studied with regard to various indices of immune system function. Although these residents had no greater incidence of infection or other general manifestations of deficiencies of the immune system, when they were challenged with specific immune tests a decrement in immune function was observed. Individuals that were exposed to the TCDD-contaminated soil failed more frequently to respond to standard antigens applied as a skin test than individuals which had not had opportunity for the TCDD exposure. In other words, there was a considerably greater incidence of anergy or relative anergy, indicating a reduced immune system responsiveness in the TCDD exposed population. Other key cell function tests also suggested a subtle effect of TCDD on the immune system.

These investigators concluded that long term exposure to 2,3,7,8 tetrachlorodibenzo-p-dioxin is associated with depressed cell mediated immunity, although the effects have not thus far resulted in excess of clinical illness in the exposed group. Additional studies will be needed to elucidate the pathophysiology and clinical significance of these immunologic findings. Since both increased incidence of malignant tumors and immunosuppression have been noted in laboratory animals exposed to TCDD, there have been frequent speculations that the immunosuppressant activity of TCDD may contribute to the mechanism of TCDD's induction of multiple types of tumors.

EPIDEMIOLOGICAL STUDIES OF CARCINOGENICITY

In recent years, the chlorophenoxy herbicides or their contaminants have been suspected contributors to increased incidence of soft-tissue sarcomas or malignant lymphomas in groups of workers who have opportunity for occupational exposure to the chlorophenoxy herbicides. Several of the observations that raised this concern are shown in the table below.

Interest in this topic was stimulated largely by studies, in the late 1970's, of an apparent association with an increased incidence of soft tissue sarcoma and exposure to herbicides or related compounds. Further interest in this subject was stimulated by epidemiological studies which found that farming and forestry occupations, particularly, had increased incidence of certain types of soft tissue sarcomas or lymphomas. Clearly, other possible etiologic factors for the increased risk of farmers to non-Hodgkins lymphoma, as reported by Cantor (1982) exists. However, a recent study by Hoar et al., (1986) indicates that increased non-Hodgkins lymphoma in Kansas farmers was associated with their opportunity for exposure to agricultural herbicides, particularly to the chlorophenoxy herbicides. The relative risk of non-Hodgkins lymphoma increased significantly with the number of days of herbicide exposure per year and with the duration or latency of exposure to herbicides. Farmers who had been exposed to heribicides more than 20 days per year, had a six-fold increased risk of non-Hodgkins lymphoma relative to non-farmers. Frequent users who mixed or applied the herbicides themselves had an increased resk of about eight-fold over the incidence of non-Hodgkins lymphoma in non-farmers. Possible exposure to tetrachlorodibenzo-p-dioxin in the use of herbicides and as a contributor to the increased non-Hodgkins lymphoma could not be ruled out. However, currently, 2,4-D is the most frequently used herbicide by farmers in the wheat

Some Selected Studies of Possible Herbicide-Related Cancer Risk

Studies (Case Control)	Odds Ratio or Relative Risk (95% CI)	Reported
Soft Tissue Sarcoma and Exposure to Phenoxacetic Acids or Chlorophenols in Sweden	5.7 (2.9-11.3)	Hardell and Sandstrom, 1979
Malignant Lymphoma and Exposure to Phenoxy Acids, or Chlorophenols and Solvents	8.5 (4,2-17.2) -combined exposures	Hardell,etal 1981
Non-Hodgkins Lymphoma in Wisconsin Farmers	1.22 (0.98-1.51) -all farmers 1.67 (1.14-2.5) ‹65 yrs. old 2.13 (1.11-4.1) -died '74-'76	Cantor, 1982
Soft Tissue Sarcomas in Agriculture and Forestry workers in England and Wales	1.7 (1.00-2.88)	Balarajan and Acheson, 1984
Malignant Lymphoma and Multiple Myeloma (MM) New Zealand Agric/Forestry workers	1.45 (1.08-1.95) -‹65 yrs. old 1.76 (1.03.3.02) -for NHL 2.22 (1.35-3.75) -for MM	Pearce, etal 1985
Non-Hodgkins Lymphoma and Agricultural Herbicide use in Kansas Farmers.	1.6 (0.9-2.6) 6.0 (1.9-19.5) >20 da/yr	Hoar, etal 1986

growing areas and it has not been shown to be contaminated with TCDD. The authors of this report also point out that the origins of non-Hodgkins lymphoma in the general population are largely unknown, but that at least one factor is immune-altering conditions and drugs. A recent report by Woods et al., (1987) supports the association between exposure to phenoxy herbicides and increased risk for non-Hodgkins lymphoma, although other interactive factors are also suggested as possible contributors.

Several of the epidemiological studies reporting possible increased incidence of tumors associated with occupational exposure to chlorophenoxy herbicides have been criticized on the basis that each of these studies, individually, often involve a very small number of cases. However, over the past few years there have been numerous independent reports of such incidents, and taken together they suggest reason for concern. The recent studies of Kansas farmers which suggests an association of non-Hodgkins lymphoma with exposure to chlorophenoxy heribicides (2,4-D remaining the major one in use) certainly indicates that the use of these herbicides should only be undertaken with full knowledge of potential hazards and full capability of preventing inhalation or dermal exposures.

REFERENCES

Balarajan, R., and Acheson, E.D. Soft tissue sarcomas in agriculture and forest workers. J Epi and Comm Hlth 38:113-116, (1984).

Bauer, H., Shulz, K.H., and Spiegelberg, U. Berufliche vergif thingen bei der herstellung von chlorphenol-verbindungen. Arch Gewerbepathol Gewerbehyg 18:538-555, (1961).

Berkeley, M., and Magee, K. Neuropathy following exposure to a dimethyline salt of 2,4-D. Arch Int Med 111:351-353, (1963).

Berwick, P. 2,4-Dichlorophenoxyacetic acid poisoning in man. J Amer Med Assoc 214:1114-117, (1970).

Blakeley, B.R., and Shiefer, B.H. The Effect of Topically Applied n-butylester of 2,4-Dichlorphenoxyacetic Acid on the Immune Response in Mice, J Appl Toxicol 6:291-295,(1986).

Bucher, N.L.R. Effects of 2,4-dichlorophenoxyacetic acid on experimental animals. Proc. Soc. Exp. Biol. Med. 63, 204-205, (1946).

Cantor, K.P. Farming and mortality from non-Hodgkins lymphoma: a case control study. Int J Cancer 29:239-242, (1982).

Courtney, K.D., and Moore, J.A. Teratology studies with 2,4,5-trichlorophenoxyacetic acid and 2,3,7,8-tetrachlorodibenzo-p-dioxin. Toxicol Appl Pharmacol 20:395-403, (1971).

Crump, K.S., and Co. Inc. Worst-Case Analysis Study on Forest Plantation Herbicide Use. Prepared for Forest Land Management Division, Department of Natural Resources, State of Washington, May 1986, (1986).

Desi, I., Sos, J., Sule, F., and Markus, V. Nervous system effects of a chemical herbicide. Arch Environ Health 4:101-108, (1962).

Donovan, J.W., Adena, M.A., Rose, G., and Battistutta, D. Case-Control Study of Congenital Anomalies and Vietnam Service (Birth Defects Study). Australian Government PublishingService, Canberra, 127pp, (1983).

Elo, H.A., and Ylitalo, P. Distribution of 2-methyl-4-chlorophenoxyacetic acid and 2, 4-dichlorophenoxyacetic acid in male rats: Evidence for the involvement of the central nervous system in their toxicity. Toxicol Appl Pharmocol 51:439-44, (1979).

Erickson, J.D., Mulinare, J., McClain, P.W., Fitch, T.G., James, L.M., McClearn, A.B., and Adams, M.J. Jr. Vietnam Veterans' Risk for Fathering Babies with Birth Defects. U. S. Deptartment of Health and Human Services, CDC Center for Environmental Health,

Atlanta, GA. August 1984, 370 pp, (1984).

Gehring, P.J., Kramer, C.G., Schwetz, B.A., Rose, J.Q., and Rowe, V.K. The Fate of 2,4,5-trichlorophenoxyacetic acid (2,4,5-T) following oral administration to man. Toxicol Appl Pharmacol 26:352-361, (1973).

Goldman, P.J. Schwerste akute Chloracne durch Trichlorophenol-zersetzungsprodukte. Arbeitsmed. Sozialmed. Arbeitshyg. 7, 12-18, (1972).

Goldstein, N., Jones, P., and Brown, J. Peripheral neuropathy after exposure to an ester of dichlorophenoxy acitic acid. J Amer Med Assoc 171:1306-1309, (1959).

Hamify, J.A., Metcalf, P., Nobbs, C.L., and Worsley, K.J. Aerial spraying of 2,4,5-T and human birth malformations: an epidemiological investigation. Science 212:349-351, (1981).

Hardell, L., Eriksson, M., Lenner, P., and Lundgren, E. Malignant lymphoma and exposure to chemicals, especially organic solvents, chlorophenols, and phenoxy acids: a case control study. Br J Cancer 43:169-176, (1981).

Hardell, L., and Sandstrom, A. Case-control study: soft -tissue sarcomas and exposure to phenoxyacetic acids or chlorophenols. Br J Cancer 39:711-717, (1979).
MD, pp. 520-536, (1982).

Hayes, W.J. Jr., Herbicides, Chapter 11 in Pesticides Studied in Man, Williams & Wilkins, Baltimore. pp. 520-536. (1982).

Hill, E.V. and Carlisle, H. Toxicity of 2, 4-dichlorophenoxyacetic acid for experimental animals. J Ind Hyg Toxicol 29:85-95, (1947).

Hoar, S.K., Blair, A., Holmes, F.F., Boysen, C.D., Rohel, R. R., Hoover, R., and Fraumeni, J.F. Agricultural herbicide use and risk of lymphoma and soft-tissue sarcoma. J Amer Med Assn, pp. 256, 1141-1147, (1986).

Hoffman, R.E., Stehr-Green, P.A., Webb, K.B., Evans, R.G., Knutsen, A.P., Schramm, W. F., Stoake, J.L., Gibson, B.B., and Steinberg, K.K. Health effects of long-term exposure to 2, 3, 7, 8-tetrachlorodibenzo-p-dioxin. J Amer Med Assn, 255:2031-2038, (1986).

Homberger, E., Reggiani, G., Sambeth, J., and Wipf, H.K. The seveso accident: its nature, extent and consequences. Ann Occup Hyg 22:327-367, (1979).

IPCS. 2, 4-Dichlorophenoxacetic acid (2, 4-D). Environmental Health Criteria 29. International Program on Chemical Safety (IPCS). World Health Organization, Geneva, Switzerland, 151 pp. (1984).

IARC. Some Fumigants, the Herbicides 2,4-D and 2,4,5-T, Chlorinated Dibenzodioxins, and Miscellaneous Industrial Chemicals. IARC monograph on the evaluation of the carcinogenic risk of chemicals to man. Volume 15. International Agency for Research on Cancer. Lyon, France, 354pp, (1977).

Khera, K.S. and McKinley, W.P. Pre- and postnatal studies on 2, 4,

5-trichlorophenoxyacetic acid, 2,4-dichlorophenoxyacetic and their derivatives in rats. Toxicol Appl Pharmacol 22:14-18, (1972).

Lamb, J.C., IV, Moore, J. A., Marks, T.A., etal. Development and viability of offspring of male mice treated with chlorinated phenoxy acids and 2,3,7,8-tetrachlorodibenzo-p-dioxin. J Toxicol Environ Health 8:835-844, (1981).

Murphy, S.D. The toxicity of pesticides and their metabolites in Degradation of Synthetic Organic Molecules in the Biosphere. Proceedings of a Conference of the National Academy of Sciences. NAS Press, Washington, D.C. pp 313-335, (1972).

Panel on Herbicides. Report on 2,4,5-T: A report of The Panel on Herbicides of the President's Science Advisory Committee. Executive Office of the President, Office of Science and Technology, U.S. Government Printing Office, Washington, D.C., (1971).

Pazderova-Vijlupkova, J., Nemcova, M., Pickova, J., Jiresek, L., and Lukas, E. The development and prognosis of chronic intoxication by tetrachlorodibenzo-p-dioxin in men. Arch Environ Health 36:5-11, (1981).

Piper, W.N., Rose, J.Q., Leng, M.L., and Gehring, P.J. The fate of 2,4,5-trichlorophenoxyacetic acid (2,4,5-T) following oral administration to rats and dogs. Toxicol Appl Pharmacol 26:339-351, (1973).

Pearce, N.E., Smith, A.H., and Fisher, D.O. Malignant lymphoma and multiple myeloma linked with agricultural applications in a New Zealand cancer registry-based study. Amer J Epidemiol 121:225-236, (1985).

Poland, A.P., Smith, D., Metter, G., and Possick, P. A health survey of workers in a 2, 4-D and 2,4,5-T plant with special attention to chloracne, porphyria cutanea tarda, and psychologic parameters. Arch Environ Health 22:316-327, (1971).

Schwetz, B.A., Sparschu, G.L., and Gehring, P.J. The effect of 2, 4-dichlorophenoxyacetic acid (2,4-D) and esters of 2,4-D on rat embryonal, fetal and neonatal growth and development. Food Cosmet Toxicol 9:801-817, (1971).

Singer, R., Moses, M., Valcirskas, J., Lilis, R., and Selikoff, I.J. Nerve conduction velocity studies of workers employed in the manufacture of phenoxy herbicides. Environ Res 29: 297-311, (1982).

Suskind, R.R. Reports on Clinical and Environmental Surveys. Monsanto Chemical Co., Nitro, West Virginia. Department of Environmental Health, University of Cincinnati College of Medicine (1953).

Todd, R. A case study of 2, 4-D intoxication. J Iowa Med Soc 52:663-664, (1962).

Vos, J.G., Krajnc, E.I., Beekhof, P.K., and van Logten, M.J. Methods for testing immune effects of toxic chemicals: evaluation of immunotoxicity of various pesticides in the rat. Miyamoto, J., and Kearney, P.C., eds. Pesticide Chemistry, Human Welfare and the Environment. Vol. 3, Mode of Action, Metabolism and Toxicology. Pergamon Press, Oxford, pp497-504, (1983).

Vos, J.G., Moore, J.A., Zinkl, J.G., Toxicity of 2,3,7,8-tetrachlorodibenzo-p-dioxin (TCDD) in C57 Bl/6 mice. Toxicol Appl Pharmacol 29:229-241, (1974).

Wallis, W.E., vanPoznak, A., and Plum, F. Generalized muscle stiffness, fasciculations, and myokymia of peripheral nerve origin. Arch Neurol 22:430-439, (1970).

Way, J.M. Toxicity and hazards to man, domestic animals, and wildlife from some commonly used auxin herbicides. Residue Rev. 26:37-62, (1969).

Woods, J.S., Polissar, L., Severson, R.K., and Heuser, M.A. Soft tissue sarcoma and non-Hodgkins lymphoma in relation to phenoxy herbicide and chorinated phenol exposure in Western Washington. J Nat Cancer Inst. in press, (1987).

Reactive Oxygen Species and Cell Injury: Role in Pesticide Toxicity

A.P. Autor
Department of Pathology
The University of British Columbia
Vancouver
British Columbia
CANADA   V6T 1W5

Reduction of Oxygen

Oxygen is metabolized by aerobic organisms in several ways including oxidative phosphorylation and mixed function oxidation. In the former process, substrates are oxidized and the substrate electrons released are transferred in discrete catalytic steps to oxygen, the terminal electron acceptor. Water is formed by this direct four electron reduction of oxygen. The energy released by the homolytic cleavage of the oxygen molecule is captured and stored as ATP. Mixed function oxidation is the process catalyzed by heme proteins which hydroxylates endogenous organic molecules to produce steroid hormones. This process also converts xenobiotics, including drugs and toxic agents, to products that are more soluble and more easily conjugated in order to facilitate elimination. In general, these processes of electron transfer are tightly controlled so that oxygen is either completely reduced to water or is incorporated into organic molecules through the catalytic action of metal-containing enzymes. Because of its electronic configuration as a diradical, however, oxygen is more easily reduced to water by sequential rather than concerted addition of four electrons. Univalent reduction leads to reactive intermediates and occurs to a small extent during normal aerobic metabolism (Fig. 1).

NATO ASI Series, Vol. H13
Toxicology of Pesticides: Experimental, Clinical
and Regulatory Aspects. Edited by L. G. Costa et al.
© Springer-Verlag Berlin Heidelberg 1987

$$O_2 + \in \longrightarrow O_2^{-\cdot} \text{ (superoxide anion free radical)} \quad (1)$$

$$O_2^{-\cdot} + 2H^+ + \in \longrightarrow H_2O_2 \text{ (hydrogen peroxide)} \quad (2)$$

$$2O_2^{-\cdot} + 2H^+ \longrightarrow H_2O_2 + O_2 \quad (3)$$

$$H_2O_2 + \in \longrightarrow OH^- + HO^{\cdot} \text{ (hydroxyl free radical)} \quad (4)$$
$$+$$
$$H^+$$
$$H_2O$$

Figure 1. Reactions are depicted at physiologic pH

Two of the three intermediates, namely $O_2^{-\cdot}$ and $HO^{\cdot}$, are free radicals; that is, reactive species which have one unpaired electron in the bonding orbital. Of the two, however, $HO^{\cdot}$ is several orders of magnitude more reactive than $O_2^{-\cdot}$ which primarily reacts with itself in a dismutation reaction to produce H_2O_2 (Reaction 3). $HO^{\cdot}$ is a powerful and undiscriminating oxidizing agent that reacts with a wide variety of organic compounds, including cellular constituents, at rate constants in the range of 1×10^9 to 1×10^{10} $M^{-1}s^{-1}$ (1). $HO^{\cdot}$-mediated oxidation of organic molecules proceeds via abstraction of an electron by $HO^{\cdot}$ thus producing an organic free radical followed by further generation of free radical moieties in succession. These two steps, typical of free radical reactions, are termed initiation and propagation, respectively. Reaction of the organic radicals with molecular oxygen produces oxygen-centered organic radicals (2,3). Typical reactions are illustrated below (Fig. 2).

Initiation:

$$RH + HO^{\cdot} \longrightarrow R^{\cdot} + H_2O \qquad\qquad (5)$$

Propagation:

$$R^{\cdot} + R'H \longrightarrow R'^{\cdot} + RH \qquad\qquad (6)$$

$$R^{\cdot} + O_2 \longrightarrow ROO^{\cdot} \text{ (peroxy radical)} \qquad (7)$$

Termination:

$$2R^{\cdot} \longrightarrow RR \qquad\qquad (8)$$

$$R^{\cdot} + XH \longrightarrow RH + Y \qquad\qquad (9)$$

$$ROO^{\cdot} + R^{\cdot} \longrightarrow ROOR \qquad\qquad (10)$$

$$ROO^{\cdot} + XH \longrightarrow ROOH + Y \qquad\qquad (11)$$

Figure 2. Free radical reactions with reference to lipid peroxidation

Free radicals vary in their stability and reactivity. As mentioned above, $HO^{\cdot}$ is extremely reactive. $O_2^{-\cdot}$, however, is much less reactive. Furthermore, some organic free radicals are long enough lived to provide the basis of molecular identification by electron spin resonance spectroscopy (4). Despite this variability in stability, the partially reduced intermediates of oxygen metabolism and their related oxygen-centered organic radicals are frequently referred to as "reactive oxygen species".

The toxic potential of oxygen was first attributed to oxygen free radicals by Gerschman who observed similarities in the pathobiology of toxic responses to hyperoxia and X-radiation. Based on the knowledge that X-radiation in aqueous solution produces oxygen radicals, Gershman formulated the unique proposal in 1954 that the oxidizing effects of oxygen free radicals are responsible for oxygen poisoning (5). Gershman termed exposure to elevated toxic concentrations of oxygen, hyperoxic stress. In an analogous manner, exposure to reactive oxygen species whatever the source is now referred to as oxidative stress.

Sources of Reactive Oxygen

It is clear now that aerobic cells are subjected to oxidative stress even under conditions of normal metabolism. In liver cells, the steady state concentration of H_2O_2 has been determined to be $10^{-9}M$ (6). Cytosolic H_2O_2 is the product of $O_2^{-\cdot}$ dismutation (Fig. 1, Reaction 3), thus after accounting for catalytic removal of $O_2^{-\cdot}$ (see below) the steady state concentration of $O_2^{-\cdot}$ in liver cells has been calculated as $10^{-12}M$ (7). Intracellular $O_2^{-\cdot}$ derives in part from the imperfect coupling of mitochondrial electron transport that allows a small but steady rate of autoxidation of certain reduced components in the chain. Reduced ubiquinone in the mitochondrial Q-cycle shows the greatest propensity for autoxidation (8).

Metabolic or environmental disturbances can provoke an enhanced rate of generation of reactive oxygen species. As a consquence of the metabolism of xenobiotics, free radicals are generated as either a major product or an accompaying by-product. The earliest known example of an organic free radical metabolite is the trichloromethyl radical ($^\cdot CCl_3$) produced by the catalytic action of cytochrome P_{450} on carbon tetrachloride (9). The reactions shown in Figure 2 illustrate the consequence of mixed function oxidation of carbon tetrachloride in the liver substituting the $^\cdot CCl_3$ that is produced catalytically for the $R^\cdot$ produced by hydrogen atom ($H^\cdot$) abstraction.

Metabolism of quinone compounds is known to consume oxygen and produce H_2O_2. The well-described toxic effects of primaquine upon erythrocytes is preceded by H_2O_2 accumulation (10). NADPH- and NADH-dependent flavoproteins catalyze the reduction of quinones to free radical semiquinones (9). Specific semiquinones autoxidize to produce reactive oxygen as follows:

$$Q + \epsilon + H^+ \longrightarrow HQ^\cdot \qquad\qquad (12)$$

$$HQ^\cdot + O_2 \longrightarrow Q + O_2^{-\cdot} + H^+ \qquad\qquad (13)$$

Oxidation of hydroquinones also produces a semiquinone intermediate:

$$H_2Q \xrightarrow[-H^+]{-e} HQ\cdot \tag{14}$$

Toxicity of quinones has been related to both the organic free radical moiety and oxidative stress from the generation of reactive oxygen intermediates. The dominant mechanism is dependent upon the chemical structure of the quinone and the extent of reactivity of its semiquinone with oxygen. Several of the quinone-containing anti-tumor antibiotics such as adriamycin (doxorubicin) express toxicity via the generation of reactive oxygen (11, 12).

Oxygen is required for the herbicidal and toxic action of paraquat (13, 14). This and other bipyridylium compounds are converted to the equivalent cation free radical by one electron reduction catalyzed by NADPH-dependent microsomal flavoenzymes. The paraquat cation radical readily transfers this electron to oxygen producing $O_2^{-\cdot}$ thus returning to the oxidized form. Continuous redox cycling is the basis of the toxicity of paraquat manifested by a constant supply of reactive oxygen species and the ultimate depletion of NADPH.

Stimulated phagocytic cells, e.g. neutrophils, produce substantial quantities of $O_2^{-\cdot}$ through univalent reduction of oxygen catalyzed by an NADPH-dependent enzyme complex located in the plasma membrane of these cells. The $O_2^{-\cdot}$ and H_2O_2 derived from the reaction are known to be required for the bactericidal action of neutrophils and are implicated in host injury associated with inflammation (15, 16).

Many tissues and organs contain xanthine dehydrogenase (XD) which catalyzes the oxidation of hypoxanthine and xanthine to xanthine and uric acid, respectively. Substrate electrons are transferred to NAD to generate NADH. Immunocytochemical analysis has localized this enzyme in the vascular endothelium (17). When the flow of oxygenated blood through the vascular system in a specific organ is blocked (ischemia), XD undergoes limited proteolysis to become xanthine oxidase (XO). Substrate to product conversion is unchanged but XO can no longer transfer electrons to NAD. Oxygen becomes the terminal electron acceptor (18) thus producing $O_2^{-\cdot}$ and H_2O_2 (19). Post-ischemic, reperfused vascular endothelium reexposed to oxygenated blood becomes a source of reactive oxygen species, therefore (20). Oxidative stress via the catalytic action of XA is strongly implicated in post-ischemic reflow injury.

Metal Catalyzed Activation of Oxygen

As described above, the "three electron reduction product" of oxygen, $HO^{\cdot}$, is an extremely powerful oxidizing agent strongly implicated as the ultimate cytotoxic agent in oxidative stress in biological systems (Fig. 3). In solution chemistry, $HO^{\cdot}$ is generated through the catalytic decomposition of H_2O_2 by ferrous iron in a reaction known as the Fenton Reaction (21) (cf. Reaction 16). In biological systems several lines of evidence identify $O_2^{-\cdot}$ and H_2O_2, both produced enzymatically, as the reactants necessary for $HO^{\cdot}$ generation. Three important qualifying factors must be applied, however: 1) The reaction requires catalysis by transition metals, commonly iron and less frequently copper 2) in biological systems iron is complexed either to small molecular weight cell constituents or to proteins in the form of non-heme iron proteins or heme proteins and 3) $HO^{\cdot}$ may not actually be generated in this form in vivo but may be liganded to iron in a complex with a redox potential equivalent to free $HO^{\cdot}$.

$O_2^{-\cdot}$ generated in biological systems may play two roles: 1) as a reductant for oxidized iron in the free or complexed form (Reaction 15) and 2) a source of H_2O_2 (Reaction 3). In vivo either extracellularly or intracellularly, the net reaction (Reaction 17) would require the redox cycling of an iron-containing compound, the precise identity of which is still unknown (22). Several intracellular candidates have been proposed including ferritin, transferrin and lactoferrin. When subjected to in vitro testing including analysis of the $HO^{\cdot}$ radical by electron spin resonance spectroscopy neither transferrin or lactoferrin, however, proved to be effective catalysts for the $O_2^{-\cdot}$-driven decomposition of H_2O_2 to $HO^{\cdot}$ (23).

$$Fe^{3+}(X) + O_2^{-\cdot} \longrightarrow Fe^{2+}(X) + O_2 \tag{15}$$

$$Fe^{2+}(X) + H_2O_2 \longrightarrow Fe^{3+}(X) + HO^{\cdot} + OH^{-} \tag{16}$$

$$O_2^{-\cdot} + H_2O_2 \xrightarrow[\text{catalyst}]{\text{iron}} HO^{\cdot} + OH^{-} + O_2 \qquad \text{Net Reaction} \tag{17}$$

Figure 3.   Iron catalysis of oxygen reduction

Strong support for the iron-mediated generation of the cytotoxic species of oxidative stress has been collected from many different laboratories by the use of iron chelators as protective agents. The most effective iron-chelating, protective agent is deferoxamine (desferrioxamine) (24). Used in the form of desferal to treat iron overload toxicity, deferoxamine is effective in preventing cytotoxicity in systems ranging from whole animal studies to cells in culture exposed to a variety of sources of oxidative stress from xanthine oxidase/xanthine to activated neutrophils (25, 26). Deferoxamine acts by preferentially chelating iron as Fe^{+3}. Fe^{+3}-deferoxamine is very difficult to reduce even by organic radicals such as the paraquat cation radical known to be an effective reductant of iron (22). When chelated in the Fe^{+3} form and incapable therefore of redox cycling, iron is effectively removed as a catalyst for $HO^{\cdot}$ production when $O_2^{-\cdot}$ and H_2O_2 are present.

In Vivo Protective Systems Against Oxidative Stress

Because the threat of oxidative stress is everpresent in oxygen metabolizing organisms, cells use enzymatic and non-enzymatic components to protect cell constituents against damage from oxidizing agents by removing or quenching reactive species. The many cell constituents which bind iron thus removing the metal as a catalyst for the decomposition of hydrogen peroxide are viewed as protective agents. These include transferrin and lactoferrin, described above. Water soluble and lipid-soluble antioxidant vitamins react with oxidizing free radicals to quench further radical reactions without producing additional free radicals. This process described as "termination" in free radical reactions. Ascorbate, α-tocopherol, β-carotene and the retinols donate electrons to quench free radicals and, although oxidized, do not become free radicals themselves. This is illustrated in Fig. 2 (Equations 9 and 11) in which the antioxidant molecule X is converted to the oxidized product Y without the generation of further free radicals.

Several enzymes and enzyme systems evolved in aerobic cells which accomplish the same free radical chain termination by catalytic means. The superoxide dismutases (SOD), metal-containing enzymes of two main structural types, catalyze the dismutation of $O_2^{-\cdot}$ (Figure 1, Reaction 3) at a rate 1×10^4 times greater than the uncatalyzed rate. The Cu/Zn-containing enzyme located in the cytosol of eukaryotes effectively lowers the intracellular steady state concentration of $O_2^{-\cdot}$. Without the catalytic action of SOD, the concentration of $O_2^{-\cdot}$ would be several orders of magnitude higher than 1×10^{-12} M^{-1} sec^{-1} (See page 1 of this chapter) calculated for the normal metabolic state. Cell concentrations of $O_2^{-\cdot}$ could be expected to be even greater under conditions of oxidative stress. For example, concomitant with the metabolism of drugs or xenobiotics such as the quinone, menadione (Vitamin K_3), or the herbicide, paraquat, the rate of $O_2^{-\cdot}$ generation is greatly elevated. SOD is necessary therefore to reduce the cytosolic concentration of $O_2^{-\cdot}$ under these conditions. Mitochondria contain a manganese form of SOD which performs the same function in the mitochondrial matrix.

Catalase removes H_2O_2 by dismuting two molecules to produce H_2O and O_2. Because catalase is packaged in the peroxisome, an intracellular compartment which also contains several peroxidases that generate H_2O_2, it is not an important agent for removal of cytosolic H_2O_2. The glutathione peroxidase system (GPX) provides the major mechanism by which intracellular H_2O_2 is removed (27). GPX reduces H_2O_2 and organic hydroperoxides (ROOH) to water or the corresponding alcohol in a reaction dependent upon reduced glutathione (GSH) as the electron source. Oxidized glutathione or glutathione disulfide (GSSG) is formed. GSSG has been used as a cellular marker for oxidative stress because elevation of the extracellular concentration of GSSG indicates an increased turnover of this system (28). The reaction is as follows:

$$H_2O_2 \ (ROOH) + 2 \ GSH \longrightarrow 2H_2O \ (ROH + H_2O) + GSSG \tag{18}$$

GSSG can be reduced back to GSH through the catalytic action of glutathione reductase using NADPH (from the hexose monophosphate pathway of glucose metabolism) as the electron source. All of these enzymes comprise the GPX system.

Sites of Cellular Injury in Oxidative Stress

In order to maintain the normal structure and function of cell constituents, all of which are vulnerable to oxidative destruction, oxidative processes are under tight control, as explained earlier in this chapter. In addition, cells and extra-cellular fluids contain many different large and small molecular weight antioxidants. Enhanced oxidative stress, however, targets many vulnerable cell components and, if unchecked, leads to cell lysis, tissue and organ necrosis and ultimately the death of the host due to organ failure.

For many years the plasma membrane has been identified as a likely target for oxidative destruction. It is composed primarily of lipids in the form of phospholipids and to a lesser extent of proteins partially or completely embedded in the phospholipid bilayer. Many of the fatty acids which comprise the phospholipid bilayer are long chain polyunsaturated acyl compounds e.g. linolenic and arachidonic acids. These fatty acids are well known targets for iron catalyzed oxidative cleavage known as lipid peroxidation (3). The first steps of lipid peroxidation follow the free radical process of initiation and oxygen insertion described in Figure 2. Continuation of the oxidation process results in the oxidative cleavage of the parent polyunsaturated fatty acid to aldehydes (alkenals) and dialdehydes (malonaldehyde) (29). Several hepatotoxins are known or suspected to express cytotoxicity through oxidative stress, that is, through the generation of reactive oxygen species.

In a series of experiments designed to examine the effects of oxidative stress on plasma membrane integrity in which cultured hepatocytes were exposed to reactive oxygen, the following results were reported 1) initial loss of cell membrane integrity is directly related to lipid peroxidation measured by the appearance of malonaldehyde, 2) deferoxamine and the antioxidant diphenylphenylenediamine (DPPD) each provide substantial protection against lipid peroxidation and cell lysis, and 3) $O_2^{-\cdot}$ drives the iron dependent decomposition of H_2O_2 leading to oxidative cell damage (30, 31).

DNA is also a vulnerable target for cell damage. Exposure of isolated fibroblasts in culture to reactive oxygen species causes strand breakage in the cellular DNA which is prevented with either catalase or iron chelators (26). Endothelial cells in culture subjected to oxidative stress from either chemical sources or activated neutrophils also show substantial single strand breaks in cellular DNA (32). Interestingly, although SOD, catalase and deferoxamine protect of endothelial cells against reactive oxygen species, the antioxidant compound DPPD which prevents lipid peroxidation is not protective (33).

Proteins are vulnerable targets for _in vitro_ oxidative damage, as well. Recently the concept has been developed that proteins are made susceptible to proteolytic degradation by oxidative modification of sensitive amino acid residues (34). Evidence has been presented that this protein modifying oxidative process is the first step in the normal turnover of proteins and has the characteristics of mixed function oxidation by the cytochrome P_{450} system. Catalase and iron chelators block proteolytic inactivation by cell extracts thus confirming the importance of oxidative modification of the target protein as a precondition to proteolysis. Originally these reactions were implicated only in normal cell protein turnover. However, several investigators have proposed that uncontrolled and non-enzymatic oxidative modification of proteins could be an important feature of cell injury and functional loss under conditions of oxidative stress (35). Such oxidative modifications could also explain the progressive degradation of extracellular matrix or structural proteins which occurs in response to chronic oxidative stress. Conditions such as emphysema might originate from such a pathophysiologic process.

Summary

A brief outline of the chemistry, biology and pathophysiology of reactive oxygen species has been presented with the purpose of applying those concepts to understanding the mechanisms of toxicity of certain pesticides. Chemicals which undergo metabolism by mixed function oxidation or are redox cycled, express toxicity partially or totally via oxidative stress. The specificity of this class of toxic agents, including pesticides, depends on several factors. These include 1) the site of metabolism which in turn is dependent upon selective transport mechanisms specific to specialized organs or cells, 2) access of reactive oxygen species to sensitive intracellular or extracellular molecular targets, and 3) intrinsic antioxidant capacity characteristic of each organ or cell type.

Morbidity and mortality resulting from exposure to xenobiotics can be better understood when the cellular and molecular changes caused by these chemicals are defined.

References

1. Anbar M and Neta, P (1967) Int J Appl Radiation and Isotopes 18:493-523

2. Mason RP (1982) In: <u>Free Radicals in Biology</u>. (ed WA Pryor) Vol 5
   Academic Press, New York pp 161-222

3. Girotti AW (1985) J Free Rad Biol Med 1:87-95

4. Janzen EG (1980) In: Free Radicals in Biology (ed WA Pryor) Vol 4
   Academic Press, New York pp 115-154

5. Gershmann R and Fenn WO (1954) Science 119:623-627

6. Chance B, Sies H and Boveris A (1979) Physiol Rev 59:527-605

7. Fridovich I (1974) Adv Enzymol 41:35-97

8. Loschen G, Azzi A, Richter C and Flohe L (1974) FEBS Lett
   42:68-72.

9. Mason RP (1979) In <u>Reviews in Biochemical Toxicology</u> (eds E. Hodgson,
   JR Bend, RM Philpot) Elsevier North Holland pp 151-200

10. Cohen G and Hochstein P (1964) Biochemistry 3:859-900

11. Rossi L, Moore GA, Orrenius S and O'Brien PJ (1986) Arch Biochem
    Biophys 251:25-35

12. Bachur NR, Gordon SL and Gee MV (1978) Canc Res 38:1745-1753

13. Homer RF, Mees GC, and Tomlinson TE (1960) J Sci Food Agri 11:309-315

14. <u>Biochemical Mechanisms of Paraquat Toxicity</u> (1977) (ed AP Autor)
    Academic Press, New York

15. Babior BM, Kipnes RS, and Curnutte JT (1973) J Clin Invest 52:741-744

16. Ward PA (1983) Adv Shock Res 10:27-34

17. Bruder G, Heid HW, Jarasch E-D, and Mather IH (1983) Differentiation
    23:218-225

18. Stirpe F, and Della Corte E (1968) J Biol Chem 244:3855-3863

19. Fridovich I (1970) J Biol Chem 245:4053-4057

20. McCord JM (1985) New Engl J Med 312:159-163

21. Haber F, and Weiss J (1934) Proc Roy Soc 147:332-351

22. Halliwell B, and Gutteridge JMC (1986) Arch Biochem Biophys
    246:501-514

23. Baldwin DA, Jenny ER, and Aisen P (1984) J Biol Chem 259:13391-13394

24. Keberle H (1964) Ann NY Acad Sci 119:758-768

25. Myers CL, Weiss SJ, Kirsch MM and Shlafer M (1985) J Mol Cell
    Cardiol 675-684

26. Mello Filho AC and Meneghini R (1984) Biochem Biophy Acta 781:56-63

27. Flohe L, Gunzler WA and Ladenstein R (1976) In <u>Glutathione:</u>
    <u>Metabolism and Function</u> (eds IM Arias and WB Jacoby) Raven Press
    New York pp 115-138

28. Kappus H and Sies H (1981) Experientia 37:1233-1240

29. Comporti M (1985) Lab Invest 53:599-623

30. Rubin R and Farber JL (1984) Arch Biochem Biophys 228:450-459

31. Starke PE and Farber JL (1985) J Biol Chem 260:10099-10104

32. Autor AP and Thies RL (1986) in Superoxide and Superoxide Dismutases
    VI (ed G Rotilio) Elsevier North Holland pp 338-342

33. Thies RL and Autor AP Unpublished data

34. Levine RL, Oliver CN, Fulks RM and Stadtman (1981) Proc Natl Acad
    Sci USA 78:2120-2124

35. Wolff SP, Garner A, and Dean RT (1986) TIBS 11:27-31

Interaction of Insecticides with the Nervous System

Lucio G. Costa
Department of Environmental Health
University of Washington
Seattle, WA 98195

Introduction

The majority of modern insecticides owe their toxicity to their ability
to attack the nervous system. Insect have a very well-developed nervous system, sometimes comparable in organization, although different and simpler,
to that of mammals (Matsumura, 1985). Poisoning of insect nervous system
easily disrupts their physiological mechanisms; most insecticides, however,
are also toxic to nontarget organisms, including human. Understanding the
mechanism of action of pesticides is a major and fundamental task for pesticide toxicologists, as the knowledge of the mechanism of action of drugs is
of vital importance for pharmacologists. Although there are cases (e.g.
aspirin) where the mechanism of action of a drug has been discovered long
after the beginning of its common and successful use, a great emphasis is
given today to address mechanistic questions in the initial stages of the
development of a new drug or pesticide. It is far beyond the scope of this
chapter to review the neurotoxicology of pesticides, and I will concentrate
only on selected aspects of mechanistic studies of pesticide neurotoxicity,
with special emphasis on their interaction with neurotransmitter receptors.
I refer the interested reader to various books and reviews published in the
last few years (Murphy, 1980; Ecobichon and Joy, 1982; Matsumura, 1985).

In general, the primary target for insecticide neurotoxicity is the
cell membrane (Doherty, 1985). Some insecticides, such as pyrethroids or
DDT, interfere with the sodium channel in the axonal membrane, thus altering
the transmembrane fluxes of Na^+. (Narahashi, 1984; Van den Bercken, this
volume; Joy, 1982). Membrane ATPases are also a common target for insecticides. The Na^+/K^+-ATPase is inhibited by DDT, chlordecone, and other organochlorine and organotin pesticides (Cutkomp et al. 1982; Desaiah, 1982;
Doherty, 1985; Costa, 1985). The Mg^{2+}-ATPase and the $(Ca^{2+}-Mg^{2+})$-ATPase,
which appears to have a relevant role in neurotransmiter release, are affected by DDT, chlordecone, and various organochlorine insecticides (Yamaguchi
et al. 1980; Cutkomp et al. 1982; Desaiah, 1982).

NATO ASI Series, Vol. H13
Toxicology of Pesticides: Experimental, Clinical
and Regulatory Aspects. Edited by L. G. Costa et al.
© Springer-Verlag Berlin Heidelberg 1987

Several insecticides, however, interfere directly with neurotransmission by affecting one or more of its components located on the cell membrane. The interaction of organophosphate and carbamate insecticides with acetylcholinesterase is certainly the most extensively studied example of such interactions. Historically, nicotine probably represents the first insecticide which exerts its toxicity by a direct interaction with a component of synaptic neurotransmission, the cholinergic nicotine receptor. The three neurotransmitter systems that have received the most attention in recent years for their interactions with insecticides, are the cholinergic, adrenergic and GABAergic systems. In particular, the interaction of various insecticides with neurotransmitter receptors has seen much progress in the past few years. This chapter will discuss the interaction of formamidine with biogenic amine receptors, those of various insecticides (organochlorines, pyrethroids, avermectin B_{1a}) with GABAergic receptors, and alteration of cholinergic receptors caused by chronic organophosphate exposure.

Adrenergic System

Formamidines, e.g. chlordimeform (CDM), its demethylated metabolite (DCDM) and amitraz, are a relatively new class of insecticides/acaricides, and the first pest control chemicals shown to exert useful effects through actions on biogenic amine systems, mainly on adrenergic and octopaminergic neurotransmission (Hollingworth and Lund, 1982). In vertebrates, norepinephrine and epinephrine have multiple essential functions both peripherally and centrally. Insects, on the other hand, are relatively deficient in norepinephrine and epinephrine (Evans, 1980a), but present a high concentration of octopamine. This amine is present only in low concentrations in the vertebrates nervous system (David and Coulon, 1985). It has been well established that, in insects, formamidines interact with the octopamine receptors and activate an adenylate cyclase (Evans and Gee, 1980; Nathanson, 1985). In insect, stimulation of adenylate cyclase and light production (in the firefly) induced by formamidines, can be antagonized by phentolamine, an alpha adrenergic blocker.

The interaction of formamidines with the biogenic amine system in mammals, however, seems to be different than in insects, and it is not yet well understood. Formamidines inhibit monoamine oxidase (MAO) activity in liver and brain both _in vitro_ and after _in vivo_ administration (Beeman and Matsumura, 1973; Benezet et al. 1976). As a result, the concentrations of serotonin, norepinephrine and dopamine in poisoned animals increase (Bailey et al. 1982). Inhibition of MAO, however, does not appear to be an important factor in the toxic effects of CDM (Robinson and Smith, 1977; Hollingworth et al. 1979; Boyes et al. 1985), suggesting that another mechanism of action should be considered. The relationship between invertebrate octopaminergic receptors and the alpha adrenergic receptors of vertebrates has been pointed out (Evans, 1980b), although differences between these two receptor systems also exist (Nathanson, 1985). Robinson and Bittle (1979) have observed that DCDM has partial agonist activity at alpha-adrenergic sites in the rabbit ear artery, and Boyes et al. (1985) have shown that the effects of CDM on rat visual function can be antagonized by the $alpha_2$-adrenergic antagonist yohimbine. Furthermore, mydriasis, bradycardia and hypertension caused by the formamidine amitraz in the rat and the dog, were specifically blocked by yohimbine (Hsu and Kakuk, 1985; Hsu et al. 1986). Yohimbine also attenuated the delayed lethality induced in mice by amitraz (Moser and MacPhail, 1985) and antagonized the antinociceptive effect of CDM (Costa, unpublished). Radioligand binding studies in mouse brain have shown that CDM, DCDM and amitraz, are potent inhibitors of the binding of [^{3}H]-clonidine to $alpha_2$-adrenoceptors (Costa and Murphy, 1986; Costa and Olibet, unpublished). In particular, the concentration of DCDM and amitraz required to inhibit 50% of the binding of [^{3}H]-clonidine was in the nanomolar range, the same range as the neurotransmitter norepinephrine, or specific $alpha_2$-adrenergic compounds such as clonidine and yohimbine. In agreement with results obtained in invertebrates and mammals indicating that $alpha_1$- or beta- adrenoceptor blockers were not capable of antagonizing the effects of formamidines (Evans and Gee, 1980; Boyes et al. 1985), CDM, DCDM and amitraz were very weak inhibitors of $alpha_1$- and beta- adrenoceptor binding in vitro (Costa and Murphy, 1986).

In summary, formamidines appear to exert their toxicity in insects through interaction with octopamine receptors; some of their effect in mammals may be mediated by interaction with $alpha_2$- adrenoceptors. On the other hand, whether their acute toxicity in mammals might be also related to such interaction, or to other effects, in particular their action on the sodium channel similar to that of local anaesthetics (Lund et al. 1976), remains to be determined.

GABAergic system

Gamma-aminobutyric acid (GABA) is regarded as one of the major inhibitory neurotransmitters in the central nervous system of vertebrates and invertebrates, and also in invertebrate skeletal muscles. There are at least three types of GABA receptors that differ in drug sensitivity, location and function (Enna and Gallagher, 1983). The $GABA_A$ receptor, located postsynaptically, is the most throughout investigated. This GABA receptor complex has a variety of drug binding sites: a site that binds GABA and other agonists as well as competitive antagonists such as bicuculline; a site that binds benzodiazepines as well as inverse agonist (e.g. beta carboline carboxylate), and a chloride channel with multiple binding sites for convulsants such as picrotoxinin, and for barbiturates (Olsen, 1981). These sites interact allosterically and affect each other's affinity for drugs. In recent years, various insecticides have been found to interact with the GABA receptor-ionophore complex and, therefore, to interfere with GABAergic transmission; these include organochlorines, pyrethroids and avermectin B_{1a}. A large class of <u>organochlorine</u> insecticides is that of the cyclodiene-type insecticides. This class of compounds include chlordane, heptachlor, aldrin, dieldrin, endrin, endosulfan, BHC and toxaphene. Despite the environmental importance of these chemicals, however, knowledge on their mechanism of action has eluded scientists for many years. Earlier work had shown that cyclodiene-type insecticides promote transmitter release at synapses in the central nervous system (Shankland, 1979). This action may be related to their ability to increase the intracellular concentration of Ca^{2+} in the presynaptic terminal possibly through their inhibition of ATPase (Yamaguchi et al. 1980). Another line of research on the mechanism of action of cyclodiene type insecticides has been the study of their interaction with the GABA receptor-ionophore complex. The initial observations arose from studies aimed at investigating the mechanism of insect resistance to cyclodiene toxicity. Matsumura and coworkers (Matsumura and Ghiasuddin, 1983; Kadous et al. 1983) found that cyclodiene resistant German cockroaches were also resistant to the lethality of a naturally occurring neuroexcitant, picrotoxinin. Picrotoxinin owes its excitatory properties to its ability to block the chloride channel and thereby antagonizing the action of the inhibitory neurotransmitter GABA (Olsen, 1981). Bicuculline also antagonizes the action of GABA, however, by a direct effect on the GABA recognition site (Enna and Gallagher, 1983). Interestingly, insects resistant to cyclodienes and picrotoxinin were not cross-resistant to bicuculline (Kaddous et al. 1983). Therefore, a modification of the picrotoxinin receptor could be responsible for the observed resistance. Using ^{3}H-labeled dihydropicrotoxinin (DHPTX), it was found that the nervous tissue of resistant insects had less binding sites for DHPTX than that of susceptible cockroaches (Kaddous et al. 1983). Further studies showed that cyclodiene-type insecticides inhibit the GABA induced increase in Cl^- permeability in the coaxal muscle of the American cockroach (Matsumura and Tanaka, 1984). Additionally, one of these compounds, heptachlor oxide, was found to inhibit the binding of DHPTX both in vitro and after in vivo administration to the cockroach (Matsumura and Tanaka, 1984). In addition to cyclodiene insecticides (aldrin, dieldrin, heptachlor, endrin, chlordane), also gamma lindane (gamma BHC) and toxaphene were capable to inhibit ^{3}H-DHPTX binding, while other chlorinated compounds (DDT, DDE, hexachlorobenzene) or other insecticides (organophosphates, carbamates, formamidines) were devoid of such activity (Tanaka et al. 1984). In this respect it is interesting to note that insect resistant to cyclodiene compounds are cross-resistant to gamma lindane, but not to DDT or organophosphates (Matsumura and Ghiasuddin, 1983).

While all these studies have been performed in insects, one of them (Matsumura and Ghiasuddin, 1983) had also shown that cyclodiene insecticides could inhibit the binding of ^{3}H-DHPTX in rat brain synaptosomes. Since signs of poisoning in rats or mice treated with gamma lindane, toxaphene and the cyclodienes are similar to those of picrotoxinin, other investigators initiated a series of studies to determine whether the GABA receptor-ionophore complex could be the target for the toxicity of these insecticides also in mammmals. A notable technical improvement of these studies was the use of ^{35}S]-TBPS (t-butylbicyclophosphorothionate) as a radioligand for the chloride channel (Lawrence and Casida, 1984). This radioligand, which derives from the potent bicyclic phosphate convulsants, has a high signal-to-noise ratio (i.e. about 70% of the binding is specific, compared with 20% of DHPTX binding; Squires et al. 1983), and allows more precise quantitative structure-activity and stereospecificity studies of possible relationships between inhibitor potencies at the TBPS (picrotoxinin) binding site and toxicity (Lawrence and Casida, 1984). Overall, the results of these experiments were similar to those observed in insects. DDT, mirex and kepone were not inhibitors of ^{35}S]-TBPS binding. Among the isomers of lindane, only the gamma inhibited the binding. Stereospecificity was demonstrated, for example, by the potency of cis-, and trans-chlordane in inhibiting ^{35}S]-TBPS binding; their IC_{50} values were 1.8 uM and 26 uM, respectively (Lawrence and Casida, 1984). In general, the potency in inhibiting the binding of [^{35}S]-TBPS binding in rat brain in vitro showed a good correlation with data of acute toxicity (Lawrence and Casida, 1984). Furthermore, a 33 and 78% decrease of [^{35}S]-TBPS binding was observed in rat brains following iv administration of a subconvulsive and convulsive dose, respectively, of dieldrin (Lawrence and Casida, 1984). In a study by Abalis et al. (1985a) similar results were obtained. These investigators also reported that cyclodiene insecticides and gamma lindane did not inhibit the binding of [^{3}H]-muscimol to GABA$_A$ receptors, nor that of [3H]flunitrazepam to benzodiazepine receptors. These results suggest that these insecticides interact specifically with a component of the GABA receptor-ionophor complex, the picrotoxinin binding site, which regulates the permeability to chloride ions. Utilizing a functional assay for GABA receptor activation, a recent study showed that insecticides of the cyclodiene-type inhibit Cl^- fluxes into rat brain membrane microsacs (Abalis et al. 1986a). The same general correlation was shown for the cyclodiene and BHC insecticides in their inhibition of [^{35}S]-TBPS binding (Lawrence and Casida, 1984; Abalis et al. 1985a; 1986a).

Another class of pesticides which interacts with the GABA receptor-ionophore complex is that of pyrethroids. These insecticides are highly toxic to insect and have generally low toxicity to mammals, forming the basis of their favorable selectivity (Casida et al. 1983). Although all pyrethroids are convulsants, they can be divided into two major classes based on neurophysiological, toxicological and pharmacological effects in a variety of species (Gammon and Sander, 1985; Lawrence et al. 1985). In mammals, Type I pyrethroids cause hyperactivity, tremor and predominantly clonic convulsions (Verschoyle and Aldridge, 1980). Representative compounds of this class are allethrin, resmethrin and permethrin. In contrast, the newer and more potent analogs containing a cyano substituent at the alpha carbon of the phenoxybenzyl alcohol moiety, have nerve effects and toxicological manifestations usually referred to as type II (Casida et al. 1983). Signs of poisoning resemble those of pycrotoxinin and include salivation, hyperactivity, choreoathetosis and clonic/tonic convulsions. Pyrethroids with Type II actions include deltamethrin, cypermethrin and fenvalerate. The target for toxic action for both classes of pyrethroids has been identified as the axonal sodium channel (Narahashi, 1984; Van den Bercken, this volume). Type II pyrethroids, however, have been found also to interact with the GABAergic system. Studies with 37 pyrethroids using rat brain synaptic membranes and

[^{35}S]-TBPS, established structure-activity relationships in vitro for inhibition of the binding of this radioligand, that are in close agreement with
the structure-toxicity relationships determined in vivo in mice (Lawrence
and Casida, 1983). Functional studies on GABA-induced chloride fluxes have
also shown that Type II pyrethroids are much more potent than Type I pyrethroids in inhibiting this effect (Abalis et al. 1986a). Different from
cyclodiene-type insecticides, inhibition of [^{35}S]-TBPS binding is not competitive (Lawrence and Casida, 1983), suggesting that their binding site is
associated with another component of the GABA receptor-ionophore complex.
Interestingly, it has been shown that cypermethrin (a Type II pyrethroid) is
about 60 times more potent as an inhibitor of [3H]-Ro5-4864 (a convulsant
benzodiazepine ligand) than of [^{35}S]-TBPS binding (Lawrence et al. 1985).
The in vitro potencies of Type II pyrethroids in the [^{3}H]-Ro5-4864 binding
assay are very close to the actual concentration in brains of poisoned animals, and also closer to the concentrations capable of activating the Na^{+}
channel (Ghiasuddin and Soderlund, 1985). Thus, one of the targets for Type
II pyrethroids on the GABA receptor-ionophore complex appears to be the
binding site for Ro5-4864, which is a selective ligand for the peripheral-
type benzodiazepine binding site. Indeed, PK 11195, an antagonist of the
peripheral-type benzodiazepine binding site, elicited a complete reversal of
the proconvulsant actions of deltamethrin and permethrin (Devaud et al.
1986). Still, kinetic studies on inhibition of [^{3}H]-Ro 5-4864 binding by
pyreth roids suggest that the binding sites for this ligand and the insecticides are very closely associated, but not identical (Lawrence et al. 1985).

An additional pesticide which interacts with the GABAergic system is
avermectin B$_{1a}$(AVM), a rather novel macrocyclic lactone disaccharide isolated from Streptomyces avermitilis, which has highly potent antihelmintic activity. AVM has been found to stimulate the release of GABA from rat brain
synaptosomes (Pong et al. 1980). There is, however, some controversy on its
effect on GABA$_{A}$ receptors. In honey bee and rat brain AVM inhibits the binding of [^{3}H]-muscimol, while in bovine brain it potentiates GABA binding
(Pong and Wang, 1982; Abalis et al. 1986b). AVM is, however, 100,000-fold
less potent in inhibiting binding of [^{3}H]-muscimol to rat brain than to honey bee brain receptors. AVM also potentiates the binding of benzodiazepines
and inhibits the binding of [^{35}S]-TBPS to the GABA-ionophor receptor complex
(Abalis et al. 1986b). This latter effect of AVM could be due to an allost-
eric interaction, since AVM did not inhibit the binding of [^{35}S]-TBPS to a
voltage-dependent chloride channel in Torpedo electric organ, a tissue that

is void of GABA$_A$ receptors (Abalis et al. 1985b). Therefore, at least in the honey bee, AVM appears to act as a GABA agonist, but more studies are needed to define its precise target in mammals.

In summary, the GABA receptor-ionophor complex, with its numerous binding sites which are reciprocally regulated, is the potential target for a variety of insecticides. Since several of these insecticides interact with other membrane components, in addition to the GABA receptor complex, one should ask which is the primary target for toxicity. For example, considering the EC$_{50}$ values in various systems it appears that the primary effects of all type of pyrethroids is on the sodium channel. This conclusion can be reached using both electrophysiological (Narahashi, 1984) and biochemical (Ghiasuddin and Soderlund, 1985) techniques. However, the finding that only type II pyrethroids can interact with the GABA receptor-ionophor complex at slightly higher concentrations (Lawrence et al. 1985), suggests that also this target can play a crucial role in the acute toxicity of this group of compounds.

Cholinergic System

Organophosphorus (OP) insecticides account for some 40% of the registered pesticides in the United States and their world market, already in the order of $1.5 billion, is projected to increase (Ecobichon, 1982). It is generally recognized that the biological activity of OP insecticides is due to their reaction with the enzyme acetylcholinesterase (AChE) and other cholinesterases. Signs and symptoms of acute poisoning with an OP insecticide are due to accumulation of acetylcholine at cholinergic synapses and may be classified into muscarinic (parasympathetic), nicotinic (sympathetic and motor) and central nervous system manifestations (Ecobichon, 1982). Of much concern are the effects of chronic, low level exposure to OPs, particularly on central nervous system functions (Costa, 1986). Several studies have described various CNS effects in workers occupationally exposed to OPs (see references in Costa, 1986). Symptoms included schizophrenic and depressive reactions, memory impairment, difficulty in concentration, irritability, anxiety, and EEG alterations. In general, however, such epidemiological studies are complicated by the lack of information on exposure, since blood cholinesterase levels are seldom reported.

A number of animal studies, on the other hand, have investigated, under more controlled experimental conditions, the effects of chronic exposure to

OPs on various parameters of the cholinergic system, from a biochemical, physiological and behavioral point of view. These studies have demonstrated that a tolerance to the toxic cholinergic effects of OPs develops upon continuous exposure (Costa et al. 1982). During a chronic feeding study with the OP insecticide parathion, Barnes and Denz (1951) noticed that rats had clearly diminished signs of toxicity after two months on the diet. This and several other studies confirmed that the development of tolerance, as evidenced by marked decreases and disappearance of a behavioral and physiological symptoms of toxicity, after chronic treatment with OPs, is a reproducible phenomenon and does not depend upon the OP used, the route of administration or the animal species (Bignami et al. 1975; Costa et al. 1982). Tolerance also developed in the absence of acute initial signs of toxicity as shown in rats repeatedly injected with a low dose of DFP or fed the insecticide disulfoton for two months (Schwab and Murphy, 1981). This latter observation suggests that tolerance might be inducible under exposure conditions that could prevail in an occupational setting. Since AChE activity was highly inhibited and acetylcholine levels were elevated in brain from OP-tolerant animals, it was suggested that tolerance permitted animals to tolerate higher concentrations of acetylcholine at neuroeffector sites. Experiments demonstrating subsensitivity of various behavioral and physiological effects of cholinergic agonists in OP-tolerant animals led to the hypothesis that a refractoriness of cholinergic receptors to acetylcholine might be involved in the development of tolerance (see references in Costa et al. 1982). Decreases in the density of muscarinic cholinergic receptors have been found in the brain and some peripheral organs of OP-tolerant animals (Costa et al. 1982). Biochemical studies on muscarinic receptor functions have shown that chronic AChE inhibition by DFP or disulfoton reduced the capacity of muscarinic agonists to inhibit adenylate cyclase activity and to stimulate hydrolysis of phosphoinositides (Olianas et al. 1984; Costa et al. 1986). These biochemical and functional alterations of cholinergic receptors are involved in the development of tolerance and subsensitivity to cholinergic agonists, although other biochemical alterations may contribute to this phenomenon (Costa et al. 1982; Gupta et al. 1985a; Raiteri et al. 1981; Russell et al. 1981). Furthermore, in studies on the development of tolerance to behavioral effects of OPs, there is evidence of a behaviorally augmented component, when animals are treated and tested on a chronic basis (Bignami et al. 1975; Giardini et al. 1982).

Tolerance might, on one hand, be considered a protective mechanism by which the organism normalizes function despite challenge from the external environment. However, the possibility that the biochemical modifications involved in this process might alter certain brain functions and/or the response to other external agents should be considered. For example, while the response to cholinergic agonists is decreased, a pharmacological super-sensitivity to antagonists has been described in chronic OP-treated animals (Russell et al. 1971). This poses the potential for an exaggerated reaction to atropine or to other compounds able to interact with the muscarinic receptor, such as antidepressants or neuroleptics. Experimental studies on the potential effects of chronic OP exposure on behavior have focussed on cognitive processes. Memory deficits in aged animals have been associated with a decreased number of brain muscarinic receptors. Since forgetfulness and memory impairment had been reported in workers chronically exposed to OPs (Metcalf and Holmes, 1969), and a decreased density of muscarinic receptors is present in brain from OP-tolerant animals (Costa et al. 1982), a few studies investigated if tolerant animals would exhibit memory impairment. Earlier studies (Reiter et al., 1973; Costa and Murphy, 1982) did not find any alteration in retention in a one-trial passive avoidance test in animals repeatedly treated with parathion and disulfoton, respectively. On the other hand, rats made tolerant to DFP were reported to have a reduced retention of a passive avoidance response as compared to control (Gardner et al. 1984). Recently, we have also found that rats exposed for two weeks to DFP or disulfoton performed significantly worse than controls in a spontaneous alternation paradigm (McDonald et al. 1986); this behavioral deficit was accompanied by an 18-28% reduction in cholinergic receptor density in cerebral cortex, hippocampus and striatum. Impairment in a passive avoidance task and decreased cortical receptor density had also been shown following a continuous infusion with physostigmine (Loullis et al. 1983).

Another subject of recurrent concern is the effects of exposure to pesticides during the pre- or postnatal period. While OPs are generally not considered to be teratogens, pre-natal exposure to low doses (doses at which no maternal toxicity is apparent) has shown various behavioral effects in the pups. These include impairment in neuromuscular strength, delays in the appearance of some reflexes, alterations in aggressive behavior, learning and locomotor activity (Spiker and Avery, 1977; Talens and Woolley, 1973). Two studies showed that prenatal exposure to DFP or methylparathion caused postnatal alterations of AChE, cholinoacetyltransferase and muscarinic receptors in developing pups (Gupta et al. 1985b; Michalek et al. 1985). In the rat brain, however, all the components of the cholinergic system are not fully developed at birth but reach adult values at one to two months. Therefore, sensitivity to an insult to the cholinergic system may be maximal during the post-natal period and lead to persistent deficits in cholinergic functions. Two studies have found that postnatal exposure of rats to DFP caused a delay in the development of cholinergic muscarinic receptors, but then the animals eventually recovered (Levy, 1981; Michalek et al. 1985). We have recently found that postnatal exposure of rats from day 5 to 21 to parathion causes a decrease in muscarinic receptor density and function at day 25 to 35, which is accompanied by a significant deficit in memory functions, measured in both a T-maze and a radial arm maze (Stamper, Balduini, Murphy and Costa, unpublished).

In summary, inhibition of AchE by OPs, particularly on a chronic basis, can lead to alterations of other parameters of the cholinergic system, which, in turn, might be associated with various behavioral disfunctions.

Conclusions

Studies on the neurotoxicity of pesticides have helped elucidating their mechanism of action in both target and nontarget species. This research responds to three general scientific needs; first, investigations on the molecular mechanism(s) of action in target species offer the possibility of designing more efficacious and selective compounds; second, this in turn, will offer explanations for unwanted effects observed in nontarget species including human, thus allowing a better understanding of the risks associated with exposure and suggestions on possible pharmacological interventions; third, this research brings important contributions to the understanding of the biological systems affected by the various pesticides, thus adding useful information to the basic sciences.

The further development of this area of research requires a combined effort of toxicologists, chemists, entomologists and epidemiologists. Furthermore, among toxicologists, an integrated approach including electrophysiologists, biochemists and behavioral toxicologists is needed, in order to build new hypotheses and abandon old ones.

Acknowledgments

The author's experimental studies were supported in part by a grant from NIEHS (ES-3424) and by a NATO grant for International Collaboration in Research. Muriel Gray and Regina Hulsman provided excellent secretarial assistance.

References

Abalis IM, Eldefrawi ME, Eldefrawi AT (1985a). High-affinity stereospecific binding of cyclodiene insecticides and gammahexachlorocyclohexane to gamma-aminobutyric acid receptors of rat brain. Pestic. Biochem. Physiol. 24: 95-102.

Abalis IM, Eldefrawi ME, Eldefrawi AT (1985b). Binding of GABA receptor channel drugs to a putative voltage-dependent chloride channel in Torpedo Electric organ. Biochem. Pharmacol. 34: 2579-2581.

Abalis IM, Eldefrawi ME, Eldefrawi, AT (1986a). Effects of insecticides on GABA-induced chloride influx into rat brain microsacs. J. Toxicol. Environ. Health 18: 13-23

Abalis IM, Eldefrawi AT, Eldefrawi ME (1986b). Action of avermectin B_{1a} on $GABA_A$ receptor and chloride channels in rat brain. J. Biochem. Toxicol. 1:69-82.

Bailey BA, Martin RJ, Downer RGH (1982). Monoamine oxidase inhibition and brain catecholamine levels in the rat following treatment with chlordimeform. Pestic. Biochem. Physiol. 17: 293-300.

Barnes JM, Denz FA (1951). The chronic toxicity of p-nitrophenyl diethyl thiophosphate (E605). J. Hyg. 49: 430-441.

Beeman RW, Matsumura F (1973). Chlordimeform: a pesticide acting upon amine regulatory mechanisms. Nature 242: 273-274.

Benezet JJ, Chang KM, Knowles CO (1976). Formamidine pesticides: metabolic aspects of neurotoxicity. IN: Pesticide and Venom Neurotoxicity (Shankpand DL, Hollingworth RM, Smyth T, eds), Plenum Press, New York, pp. 789-206.

Bignami G, Rosic N, Michalek H, Milosevic M, and Gatti GL (1975). Behavioral toxicity of anticholinesterase agents: methodological, neurochemical and neuropsychological aspects. In: Behavioral Toxicology (Weiss B, Laties VG, eds.), Plenum Press, New York, pp. 155-215.

Boyes WK, Moser VC, MacPhail RC, Dyer RS (1985). Monoamine oxidase inhibition cannot account for changes in visual evoked potentials produced by chlordimeform. Neuropharmacol. 24: 853-860.

Casida JE, Gammon DW, Glickman AH, Lawrence LJ (1983). Mechanisms of selective action of pyrethroid insecticides. Ann. Rev. Pharmacol. Toxicol. 23: 413-438.

Costa LG (1985). Inhibition of gamma-$[^3H]$ aminobutyric acid uptake by organotin compounds in vitro. Toxicol. Appl. Pharmacol. 79: 471-479.

Costa LG (1986). Organophosphorus compounds. In: Recent Advances in Nervous System Toxicology (Galli CL, Manzo L, Spencer PS, eds.), Plenum Press, New York (in press).

Costa LG, Murphy SD (1982). Passive avoidance retention in mice tolerant to the organophosphorus insecticide disulfoton. Toxicol. Appl. Pharmacol. 65: 451-458.

Costa LG, Murphy SD (1986). Interaction of the pesticide chlordimeform with adrenergic receptors in mouse brain: an in vitro study. Arch. Toxicol. (submitted).

Costa LG, Schwab BW, Murphy SD (1982). Tolerance to anticholinesterase compounds in mammals. Toxicology 25: 79-97.

Costa LG, Kaylor G, Murphy SD (1986). Carbachol- and norepinephrine- stimulated phosphoinositide metabolism in rat brain: effect of chronic cholinesterase inhibition. J. Pharmacol. Exp. Ther. (in press).

Cutkomp LK, Koch RB, Desaiah D (1982). Inhibition of ATPases by chlorinated hydrocarbons. In: Insecticide mode of action (Coats JR, ed.), Academic Press, New York, pp. 45-69.

David JC, Coulon JF (1985). Octopamine in invertebrates and vertebrates. A review. Progr. Neurobiol. 24: 141-185.

Desaiah D (1982). Biochemical mechanisms of chlordecone neurotoxicity. A review. Neurotoxicology 3: 103-110.

Devaud LL, Szot P, Murray TF (1986). PK 11195 antagonism of pyrethroid-induced proconvulsant activity. Eur. J. Pharmacol, 120: 269-273.

Doherty JD (1985). Membrane effects of pesticides. In: Neurotoxicology (Blum K, Manzo L, eds.), Marcel Dekker Inc., Basel, pp. 405-421.

Ecobichon DJ (1982). Organophosphorus ester insecticides. In: Pesticides and Neurological Diseases (Ecobichon DJ, Roy RM, eds.), CRC Press, Boca Raton, pp. 15-52.

Ecobichon DJ, Joy RM (eds.) (1982). Pesticides and Neurological Diseases. CRC Press, Boca Raton, pp. 280.

Enna SJ, Gallagher JP (1983). Biochemical and electrophysiological characteristics of mammalian GABA receptors. Int. Rev. Neurobiol. 24: 181-212.

Evans PD (1980a). Biogenic amines in the insect nervous system. Adv. Insect Physiol. 15: 317-473.

Evans PD (1980b). Octopamine receptors in insects. In: Receptors for Neurotransmitters, Hormones and Pheromones in Insects (Sattelle DB, Hall LM, Hildebrand JG, eds.), Elsevier, New York, pp. 245-258.

Evans PD, Gee JD (1980). Action of formamidine pesticides on octopamine receptors. Nature 287: 60-62.

Gammon DW, Sander G (1985). Two mechanisms of pyrethroid action: electrophysiological and pharmacological evidence. Neurotoxicol. 6: 63-86.

Gardner R, Ray R, Trankenheim J, Wallace K, Loss M, Robichand R (1984). A possible mechanism for diisopropylfluorophosphate-induced memory loss in rats. Pharmacol. Biochem. Behav. 21: 43-46.

Ghiasuddin SM, Soderlund DM (1985). Pyrethroid insecticides: potent stereospecific enhancers of mouse brain sodium channel activation. Pestic. Biochem. Physiol. 24: 200-206.

Giardini V, Meneguz A, Amorico L, DeAcetis L, Bignami G (1982). Behaviorally augmented tolerance during chronic cholinesterase reduction by paraoxon. Neurobehav. Toxicol. Teratol. 4: 335-345.

Gupta RC, Patterson GT, Deltbarn WD (1985a). Mechanisms involved in the development of tolerance to DFP toxicity. Fund. Appl. Toxicol. 5: 517-528.

Gupta RC, Rech RH, Lovell KL, Welsch F, Thornburg JE (1985b). Brain cholinergic, behavioral and morphological development in rats exposed in utero to methylparathion. Toxicol. Appl. Pharmacol. 77: 405-413.

Hollingworth RM, Lund AE (1982). Biological and neurotoxic effects of amidine pesticides. In: Insecticide Mode of Action (Coats JR, ed), Academic Press, New York, pp. 198-227.

Hollingworth RM, Leister J, Ghali G (1979). Mode of action of formamidine pesticides: an evaluation of monoamine oxidase as the target. Chem. Biol. Interactions 24: 35-49.

Hsu WH, Kakuk TJ (1984). Effect of Amitraz and chlordimeform on heart rate and pupil diameter in rats: mediated by alpha$_2$-adrenoceptors. Toxicol. Appl. Pharmacol. 73: 411-415.

Hsu WH, Lu ZX, Hembrough FB (1986). Effect of Amitraz on heart rate and aortic blood pressure in conscious dogs: influence of atropine, prazosin, tolazoline and yohimbine. Toxicol. Appl. Pharmacol. 84: 418–422.

Joy RM (1982). Chlorinated Hydrocarbon Insecticides. In: Pesticides and Neurological Diseases (Ecobichon DJ, Joy RM, eds.), CRC Press, Baton Rouge, pp. 91–150.

Kaddous AA, Ghiasuddin SM, Matsumura F, Scott JG, Tanaka K (1983). Difference in the picrotoxinin receptor between the cyclodiene-resistant and susceptible strains of the german cockroach. Pestic. Biochem. Physiol. 19: 157–166.

Lawrence LJ, Casida JE (1983). Stereospecific action of pyrethroid insecticides on the gamma-aminobutyric acid receptor-ionophore complex. Science 221: 1399–1401.

Lawrence LJ, Casida JE (1984). Interactions of lindane, toxaphene and cyclodienes with brain-specific t-butylbicyclophosphorothionate receptors. Life Sci. 35: 171–178.

Lawrence LJ, Gee KW, Yamamura HI (1985). Interations of pyrethroid insecticides with chloride ionophore-associated binding sites. Neurotoxicol. 6:87–98.

Levy A (1981). The effect of cholinesterase inhibition on the autogenesis of central muscarinic receptors. Life Sci. 29: 1065–1070.

Loullis CC, Dean RL, Lippa AS, Meyerson LR, Beer B, Bartus RT (1983). Chronic administration of cholinergic agents: effects on behavior and calmoduli. Pharmacol. Biochem. Behav. 18: 601–604.

Lund AE, Yim GKW, Shankland DL (1976). The cardiovascular toxicity of chlordimeform: a local anesthetic-like action. In: Pesticide and Venom Neurotoxicity (Shankland DL, Hollingworh RM, Smyth T, eds.), Plenum Press, New York, pp. 171–177.

Matsumura F (ed.) (1985). Toxicology of Insecticides. Plenum Press, New York, pp. 598.

Matsumura F, Ghiasuddin SM (1983). Evidence for similarities between cyclodiene type insecticides and picrotoxinin in their action mechanisms. J. Environ. Sci. Health B 18: 1–14.

Matsumura F, Tanaka K (1984). Molecular basis of neuro-excitatory actions of cyclodiene-type insecticides. In: Cellular and Molecular Neurotoxicology (Narahashi T, ed), Raven Press, New York, pp. 225–240.

Mcdonald BE, Costa LG, Murphy SD (1986). Memory impairment following repeated organophosphate exposure. Pharmacologist 28: 226.

Metcalf DR, Holmes JH (1969). EEG, psychologixal and neurological alterations in humans with organophosphorus exposure. Am. N.Y. Acad. Sci. 160: 357–365.

Michalek H, Pintor A, Fortuna S, Bisso GM (1985). Effects of diisopropylfluorophosphate on brain cholinergic systems of rats at early developmental stages. Fund. Appl. Toxicol. 5: S204–S212.

Moser VC, MacPhail RC (1985). Yohimbine attenuates the delayed lethality induced in mice by amitraz, a formamidine pesticide. Toxicol. Lett. 28: 99–104.

Murphy SD (1980). Pesticides In: Casarett and Doull's Toxicology (Doull J, Klassen CD, Amdur MO, eds.), MacMillan, New York, pp. 357–408.

Narahashi T (1984). Nerve membrane sodium channels as the target of pyrethroids. In: Cellular and Molecular Neurotoxicology (Narahashi T, ed.), Raven Press, New York, pp. 85–108.

Nathanson JA (1985). Characterization of octopamine-sensitive adenylate cyclase: elucidation of a class of potent and selective octopamine-2 receptor agonists with toxic effects in insects. Proc. Natl. Acad. Sci. USA 82: 599–603.

Olianas MC, Onali P, Schwartz JP, Neff NH, Costa E (1984). The muscarinic receptor adenylate cyclase complex of rat striatum: desensitization following chronic inhibition of acetyl- cholinesterase activity. J. Neurochem. 42: 1439-1443.

Olsen RW (1981). The GABA postsynaptic membrane receptor-ionophore complex. Mol. Cell. Biochem. 39: 261-279.

Pong SS, Wang CC (1982). Avermectin B_{1a} modulation of gammaaminobutyric acid receptors in rat brain membranes. J. Neurochem. 38: 375-379.

Pong SS, Wang CC, Fritz LC (1980). Studies on the mechanism of action of avermectin B_{1a}: stimulation of release of gammaaminobutyric acid from brain synaptosomes. J. Neurochem. 34: 351-358.

Raiteri M, Marchi M, Paudice P (1981). Adaptation of presynaptic acetylcholine autoreceptors following long-term drug treatment. Eur. J. Pharmacol. 74: 109-110.

Reiter L, Talens G, Woolley D (1973). Acute and subacute parathion treatment: effects on cholinesterase activities and learning in mice. Toxicol. Appl. Pharmacol. 25: 582-588.

Robinson CP, Smith PW (1977). Lack of involvement of monoamine oxidase inhibition in the lethality of acute poisoning by chlordimeform. J. Toxicol. Env. Health 3: 565-568.

Robinson CP, Bittle I (1979). Vascular effects of demethylchlordimeform, a metabolite of chlordimeform. Pestic. Biochem. Physiol. 11: 46-55.

Russell RW, Vasquez BJ, Overstreet DH, Dalglish FW (1971). Effects of cholinolytic agents on behavior following development of tolerance to low cholinesterase activity. Psychopharmacologia (Berl.) 20: 32-41.

Russell RW, Carson VG, Booth RA, Jenden DJ (1981) Mechanisms of tolerance to the anticholinesterase DFP: acetylcholine levels and dynamics in the rat brain. Neuropharmacol. 20: 1197-1201.

Schwab BW, Murphy SD (1981). Induction of anticholinesterase tolerance in rats with doses of disulfoton that produce no cholinergic signs. J. Toxicol. Env. Hlth. 8: 199-204.

Shankland DL (1979). Action of dieldrin and related compounds on synaptic transmission. IN: Neurotoxicology of Insecticides and Pheromones (Narahashi T, ed.), Plenum Press, New York, pp. 139-153.

Spyker JM, Avery DL (1977). Neurobehavioral effects of prenatal exposure to the organophosphate diazinon in mice. J. Toxicol. Env. Health 3: 989-1002.

Squires RF, Casida JE, Richardson M, Saederup E (1983). [^{35}S]t-Butyl bicyclophosphorothionate binds with high affinity to brain-specific sites coupled to gamma-aminobutyric acid A and ion recognition sites. Mol. Pharmacol. 23: 326-336.

Talens G, Woolley D (1973). Effects of parathion administration during gestation in the rat on development of the young. Proc. West. Pharmacol. Soc. 16: 141-145.

Tanaka K, Scott JG, Matsumura F (1984). Picrotoxinin receptor in the central nervous system of the american cockroach: its role in the action of cyclodiene type insecticides. Pestic. Biochem. Physiol. 22: 117-127.

Verschoyle RD, Aldridge WN (1980). Structure-activity relationships of some pyrethroids in rats. Arch. Toxicol. 45: 325329.

Yamaguchi I, Matsumura F, Kadous AA (1980). Heptachlor epoxide: effects on calcium-mediated transmitter release from brain synoptosomes in rat. Biochem. Pharmacol. 29: 1815-1823.

The role of genetic differences in human susceptibility to pesticides

G. S. Omenn
School of Public Health and Community Medicine
and Dana Program in Genetics and Environmental Health
University of Washington
Seattle, Washington 98195, USA

Pesticides are among the most intensively studied and most extensively regulated classes of chemicals throughout the world. In the
United States, regulatory decision-making addresses both the specific
chemical and the various applications that may lead to significant exposures. In the general framework for risk assessment and risk management, variation in susceptibility is emerging as a variable of interest,
as shown in Figure 1:

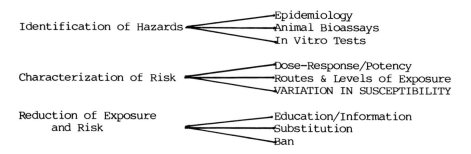

Identification of Hazards
  Epidemiology
  Animal Bioassays
  In Vitro Tests

Characterization of Risk
  Dose-Response/Potency
  Routes & Levels of Exposure
  VARIATION IN SUSCEPTIBILITY

Reduction of Exposure
and Risk
  Education/Information
  Substitution
  Ban

Figure 1: Framework for Decisions about Pesticides and Other Chemicals

Such host variation may be due to genetic traits, age, sex, diet, co-
existing exposures, pre-existing medical conditions, medications, protective measures, and lifestyle behavioral factors. In turn, genetic
factors may act through differences in metabolism (activation and inactivation) and through differences in susceptibility of sites of action for
the chemical in target tissues. Such studies of genetically-predisposed
differences to effects of environmental agents comprise the field of
"eco-genetics" (1, 2).

There is genetic variation at each of three major sites for metabolism of organophosphorus pesticides: the cytochrome P450 microsomal
mixed function oxidases; the plasma phosphotriesterases, including paraoxonase; and the glutathione-S-transferases. Representing many pesti-

cides subject to metabolism by these systems, parathion (O,O–diethyl p–nitrophenyl phosphorothionate) will serve as the main example (3).

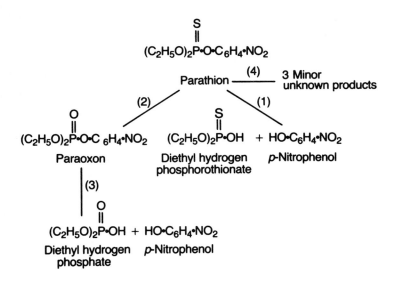

Figure 2: **Metabolism of Parathion (3):** Reactions (1) and (2) are catalyzed by different cytochrome P450 microsomal oxidases, reaction (3) by the phosphotriesterase paraoxonase, and reaction (4), in part, by glutathione S–transferase.

CYTOCHROME P450 MICROSOMAL OXIDASES

Notable genetic variation has been reported in the human and mouse cytochrome P450 systems, involving baseline enzyme activity and responsiveness to inducers. Parathion is metabolized by at least two different P450–mediated pathways to paraoxon and to diethyl phosphorothioic acid (diethyl hydrogen phosphorothionate) in liver, lung, and several other tissues. Based upon differential effects of inhibitors and activators, Neal demonstrated 20 years ago that two separate mixed–function oxidases were responsible for metabolizing parathion to paraoxon and to diethyl phosphorothioic acid + p–nitrophenol (3). Little is known about interindividual variation in these steps or about the particular cytochrome P450s responsible for parathion oxidation.

In contrast, some remarkable evidence of genetically polymorphic variation in P450 metabolism has emerged in the past decade for certain pharmaceutical agents. [By convention a variant is called polymorphic when its frequency exceeds 1 percent in the tested population.]

Debrisoquine, an unsuccessful antihypertensive drug, is the first chemical for which monogenic control of its oxidative metabolism was discovered in human population studies. Most individuals readily metabolize this drug to 4-hydroxy-debrisoquine, its major urinary metabolite. However, up to 10 percent of Caucasian populations excrete very low amounts of the 4-hydroxy-derivative; autosomal recessive inheritance is responsible. At least 20 additional drugs are subject to the same polymorphism, with 90% of people extensive metabolizers and 10% poor metabolizers. Amazingly, the defect affects oxidation of alkyl, aryl, and aromatic carbons, nitrogen, and sulfur atoms in certain substrates. Meanwhile, other drugs such as antipyrine, acetanilide, and tolbutamide are oxidized normally by the poor metabolizers of debrisoquine. Poor hydroxylators have low levels of debrisoquine-4-hydroxylase activity but normal levels of overall cytochrome P450 and normal catalytic activities toward other substrates. Barbeau et al recently compared 40 Parkinson's disease patients and 40 control subjects by debrisoquine metabolism phenotypes; they found significantly more poor or intermediate metabolizers among the patients, and among Parkinson's disease patients they found a correlation of poor or intermediate hydroxylators with earlier onset of the disease (4). They postulate that such patients might be unable to handle exposures to unidentified environmental neurotoxins (?pesticides) that would not harm extensive metabolizers. However, one must be cautious in interpreting these results, especially if some of the patients had received neuroleptic drugs, which are known to impair debrisoquine oxidation (5). Quinidine has similar inhibitory effects (W. Kalow, personal communication, 1986).

The debrisoquine polymorphism stimulated tremendous pharmacologic research activity in the cytochrome P450 system, and additional polymorphic drug-oxidizing traits have been reported. Guengerich and his colleagues have led the way in purification and immunochemical analysis of P450 isozymes from both rat and human liver. Distlerath et al have purified to homogeneity a minor form of P450 from male Sprague-Dawley rat liver which appears to be the analogue of the polymorphic debrisoquine-metabolizing P450 in humans (6). The same research group has now published the purification of human P450s involved in four distinct P450 polymorphisms, namely 4-hydroxylation of debrisoquine, O-deethylation of phenacetin, 4-hydroxylation of S-mephenytoin, and dehydrogenation of nifedipine to its pyridine metabolite (7).

Nifedipine oxidation is of special interest to us because the same polymorphic P450 seems to be responsible for epoxidation of the pesticide aldrin. Two different P450s participate in the metabolism of nifedipine in the rat, but in human liver a single isozyme seems to be responsible; it cross-reacts with the pregnenolone-inducible P450 in the rat. No sex or age dependence of nifedipine metabolism has been observed; the highest activity was found in a patient treated with dexamethasone, so glucocorticoids may be inducers of this P450. Various correlation, reconstitution, and immunochemical inhibition data indicate that other substrates for this P450 isozyme include 2- and 4-hydroxylation of 17-beta-estradiol, 6-beta-hydroxylation of testosterone, and demethylation of d-benzphetamine (see 7).

Isolation and characterization of the P450 proteins and genes is progressing rapidly. The functional roles of these P450s can be identified in vitro and in vivo. The demonstration of congruent patterns of extensive or poor metabolism for drugs affected by the debrisoquine polymorphism emerged from clinical studies in which test doses of the various drugs were given to subjects whose phenotype for debrisoquine metabolism had been determined. This approach is not useful for toxic environmental chemicals for which exposure should be avoided, such as parathion. Instead, it is necessary to rely upon analogies with metabolism in whole animals and in microsomes from animal tissues. In vitro studies of the substrate specificity of purified human P450 enzymes is becoming an alternative approach.

With regard to parathion, it is well-established that both phenobarbital and methylcholanthrene pretreatments in rabbits induce significant increases in the activity (Vmax) of liver and lung microsomes converting parathion to paraoxon and to diethyl phosphorothioic acid (8). Phenobarbital and methylcholanthrene, in turn, are known to induce quite different major P450 isozymes. Thus, parathion may be metabolized by two or more major P450s for which no inherited variation in the baseline activity level is known at this point. However, inherited variation in the extent of inducibility of increases in the activity of these P450s is well-known, especially at the Ah locus in the mouse. Ah signifies inducibility of aryl hydrocarbon hydroxylase (AHH) activity (see reference 2 for additional information). The Ah gene product appears to be a cytosolic receptor protein with high affinity (Kd = 1 nM) for polycyclic aromatic inducers; the receptor is defective in nonresponsive strains of mice. Variation at the Ah locus is associated with marked differences in susceptibility to carcinogen-induced solid tumors and leukemias and to several drug-induced organ-specific toxicities (9).

Numerous studies have assayed human peripheral lymphocytes in culture, stimulated by phyto-hemagglutinin (PHA) and then exposed to known inducers of P450s. People whose cells show high induction of AHH (which activates aryl hydrocarbon procarcinogens) have been reported to be at much higher risk of cigarette smoking-associated lung cancer and laryngeal cancer, but results are inconsistent, due to technical problems with the assays (see Kouri et al in ref 2). Early reports from Kellerman included family studies that seemed to show a single-gene effect on inducibility, with high, intermediate, and low inducers (approximately 10, 45, and 45 percent of the population). The single-gene determination seemed consistent with the Ah locus in the mouse, but it is now clear that multiple other factors affect AHH inducibility in the mouse and in humans. Despite the technical problems and the controversy in this arena, we should expect valuable assays to emerge as monoclonal anti-bodies and DNA probes are developed for specific P450s and applied to lymphocytes, and as genetic variants are found which can be tools for dissection of the substrate specificity of the multiple P450s. DNA probes will be useful to determine specific mRNA levels and to search for restriction fragment length polymorphisms of DNA in and near the cyto-chrome P450 genes (10). Such analytical tools will open the way to analyze both baseline activity and inducibility of these key enzymes in the metabolism of environmental chemicals (xenobiotics) including parathion.

As reported at the 7th International Congress on Human Genetics in Berlin two weeks before our NATO Institute, human P450 genes have been localized to chromosome 15 for the methylcholanthrene-inducible P450, chromosome 19 for the phenobarbital-inducible P450, chromosome 22 for the debrisoquine-metabolizing P450, and chromosome 6 for the P450 involved in hydroxylation of certain steroids and deficient in the inherited disorder congenital adrenal hyperplasia. Yet another P450 of considerable eco-genetic interest has been characterized in human and rabbit liver (11): a P450 induced by and/or oxidizing ethanol, trichloroethylene, benzene, aniline, acetone, pyrazole, isoniazide, imidazole, and carbon tetra-chloride. It converts acetaminophen to a toxic intermediate and de-methylates the carcinogen N-nitrosodimethylamine (11).

PLASMA PHOSPHOTRIESTERASES/PARAOXONASE

The active toxic metabolite of parathion is paraoxon. Inactivation of paraoxon is critical to protection against the toxic neurological effects. The responsible plasma enzyme, paraoxonase, varies 3-6 fold in activity among humans, with about 50% of the population having the low-activity form of the enzyme and 50% having the intermediate or high activity forms. A single autosomal recessive gene determines low-activity. In fact, it is now known that this gene is carried on chromosome 7 in humans, closely linked to the gene for cystic fibrosis.

Furlong, Motulsky, et al at the University of Washington have purified human paraoxonase from albumin esterase and contributed to the biochemical characterization of the enzyme. Studies of paraoxon toxicity in rats and rabbits in vivo and in vitro by LaDu and associates at the University of Michigan and by Costa and Furlong at the University of Washington indicate that lower plasma paraoxonase activity is, indeed, associated with greater toxicity at the same dose. Field studies are underway by Richter et al in Israel to attempt to assess paraoxonase variation in combination with variation in exposures.

Paraoxonase is an aryl-esterase capable of catalyzing the hydrolysis of additional organophosphates, including DFP, tabun, sarin, chlorpyrifos-oxon, and probably malaoxon; some carbamates, such as carbaryl and 3-isopropylphenyl-N-methyl-carbamate; and certain aromatic carboxylic acid esters, such as phenylacetate, 4-nitro-phenyl-acetate, and 2-naphthylacetate. The enzyme requires Ca^{++} for activity and is inhibited by sulfhydryl-binding agents, such as p-chloromercuribenzoate, and calcium-chelating agents such as EDTA. The primary physiological role for this plasma enzyme is not known. Quite a variety of assays have been developed to attempt to separate the three phenotypes of low, intermediate, and high activity into at least bimodal (low and intermediate/high) and often trimodal distributions (see ref 2). Some populations, especially in Africa, may have different allele frequencies, different alleles, or different modifying factors, since unimodal distributions of plasma paraoxonase activity profiles have been obtained by several research groups. LaDu et al, however, applied qualitative criteria to an initially unimodal distribution of paraoxonase activity in a Sudanese population and were able to distinguish all three phenotypes. Their findings were supported by analysis of pedigrees for 22 families (12). In their work,

1 M NaCl is used as a differential stimulating factor, increasing B isozyme 2-3 fold, while little affecting A isozyme; the ratio of para-oxonase activity to phenylacetate hydrolysis is 7 times higher for B than for A isozyme; and isozyme B has much more residual activity than A after inhibition by 0.1 mM chlorpromazine. Their explanation for the lack of readily detectable bimodality of paraoxonase activity in the Sudanese and possibly in some other Third World populations is a higher prevalence of the B or "high" activity allele (0.48 vs 0.29 for several U.S. and European populations).

Motulsky's group resolved the trimodality in Caucasian populations with an assay in the presence and absence of EDTA (13). Then gel chro-matographic methods were employed to show that paraoxonase activity eluted in two peaks, one of which is serum albumin (14). It was impos-sible to separate this paraoxonase fraction from albumin by various methods; evidence that albumin itself accounts for the paraoxonase/ester-ase activity was obtained by analysis of serum from a patient with genetic deficiency of albumin (analbuminemia), in which this second fraction of paraoxonase was absent as well. The polymorphic variation in paraoxonase activity (low, intermediate and high) involves only the non-albumin fraction, which is stimulated by 1 M NaCl and inhibited by 10 mM EDTA. It is important to note that the albumin-associated activity is detectable only at pH above pH 8, so it may have little or no _in vivo_ significance. However, most laboratory assays for paraoxonase are carried out above pH 8.

The arylesterase activity assayed with phenylacetate does not vary according to the genotype and phenotype of the individual with regard to paraoxon hydrolysis. Recent work in Motulsky's laboratory (Seidel, Costa, Richter and Furlong, unpublished), has shown that hydrolysis of chlorpyrifosoxon is similarly unimodal and is correlated with activity against phenylacetate, even though chlorpyrifos-oxon is a competitive inhibitor (as is malaoxon) of paraoxonase activity. Presumably the same enzyme protein mediates both enzyme activities (paraoxonase and aryl-esterase), but there must be different interactions of the two different classes of substrates with the active site. Furlong continues work to isolate and characterize the enzyme and its gene.

As reported at this meeting by Costa et al (15), _in vivo_ studies com-paring susceptibility of rats and rabbits to neurotoxicity from para-thion/paraoxon strongly support the conclusion that low paraoxonase activity in plasma is associated with lower LD50 for paraoxon, i.e. higher toxicity from a given dose; no comparable human data are available as yet. Also, we should note that there are data thus far about possible inter-individual variation in activity of the neurotoxic esterase.

GLUTATHIONE S-TRANSFERASES

Parathion also undergoes conjugation with glutathione (included in "minor products" in Figure 2). Glutathione S-transferases catalyze conjugation of electrophiles with reduced glutathione (GSH), leading to further metabolism and excretion as mercapturic acids. Also these enzyme proteins can act as binding proteins (previously termed "ligandin"). The multiplicity and functional properties of these enzymes have been studied in several species, notably in rat and human liver (16).

An early indication of the role of GSH S-transferases in metabolism of organophosphorus insecticides was the finding of two metabolites of dichlorvos (2,2-dichlorovinyl dimethyl phosphate), one from cleavage at the P-O-methyl bond and the other at the P-O-vinyl bond. Then Fukami's group in Japan reported that methyl parathion is demethylated by the supernatant fraction of rat liver homogenate in the presence of GSH. Both O-alkyl conjugation and O-aryl conjugation occur, and the specific conjugates have been identified (see 17). Most of the organophosphorus compounds studied as substrates for GSH S-transferase reactions have been either phosphorothionates, $(RO)_2P(S)OR'$, or phosphates, $(RO)_2P(O)OR'$, though phosphorodithioates, $(RO)_2P(S)SR'$, and phosphorothiolates, $(RO)_2P(S)SR'$, have been studied, too. Both mammalian and insect transferases have a preference for methoxy compound substrates; O-alkyl conjugation is extremely slow when the methoxy groups are replaced by ethoxy or long alkoxy groups, hence methyl parathion is a much better substrate than parathion (ethyl parathion). O-aryl conjugation predominates with ethyl parathion, while O-alkyl conjugation is greater for methyl parathion (Figure 3).

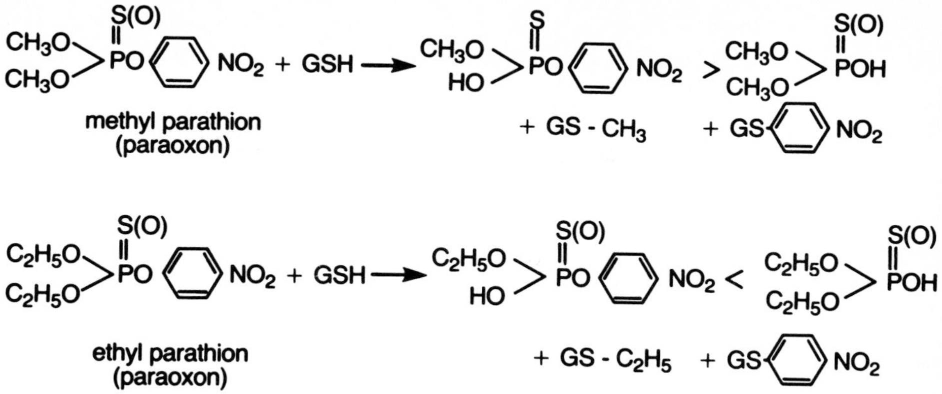

Figure 3: Conjugation of Methyl Parathion and Ethyl Parathion with Glutathione (18)

Genetic analyses are needed to distinguish whether the multiple forms of GSH S-transferases in a given tissue homogenate are due to multiple isozymes from polypeptide subunits and their post-translational modification or are products of multiple genes and different alleles at those gene loci. Both phenomena are involved in the multiple isozymes found in human liver by ion-exchange chromatography and isoelectric focusing (16).

Broad (19) utilized the technique of starch-gel electrophoresis and specific staining for GSH S-transferase activity to show genetic variation involving at least two loci in the liver. A third locus specifies the form of GSH S-transferase in erythrocytes, which had no variants in his sample populations. 1-chloro-2,4-dinitrobenzene was used as the substrate in these studies. Three different alleles were inferred at the GST-1 locus for the anodally-migrating transferases, leading to four distinguishable phenotypes on stained gels; one of the alleles was a null allele, apparently specifying an inactive product. The two active alleles specify differently-migrating subunits; since the enzyme is a dimer, three dimer possibilities are expected and are observed. Hardy-Weinberg calculations of the expected ratios of the phenotypes in the 179 liver samples fit observations when the null allele is included in the derivation. The second locus, termed GST-2, specifies transferase subunits which again give a three-phenotype pattern from two different active subunits, migrating toward the cathode at pH 8.6 in Tris-EDTA-borate buffer. However, Strange et al found differences in the phenotype of the cathodally-migrating GST-2 isozymes among tissues from single individuals, indicating that variation is probably not inherited at the GST-2 locus (20).

Recently Seidegard and Pero demonstrated a polymorphism of GSH S-transferase in blood mononuclear cells from human subjects, using tritiated trans-stilbene oxide as substrate (21). Among 248 unrelated individuals, 133 (54%) had extremely low or undetectable activity (< 25 pmol/min/10^7 cells), while 94 (38%) had activity between 600 and 1400 (intermediate) and 21 (8%) had activity between 1750 and 2200 pmol/min/10^7 cells. These frequencies for the enzyme phenotypes fit very well a single gene model with gene frequencies of 0.73 for the low activity allele and 0.27 for the very high activity allele. The pattern of inheritance of GST activity in 8 families likewise fit a single gene non-sex-linked model.

In our own group, Hornung and Motulsky have confirmed this striking variation in GSH S-transferase activity toward trans-stilbene oxide in white blood cells. Analysis of the phenotypes among 35 children in 15 families indicates a good fit to a single-gene determination of low-intermediate-high activity (unpublished). It will be interesting to test whether the trimodality of GSH-S-transferase in lymphocytes toward trans-stilbene oxide holds for the dealkylation of methylparathion. Anderson and Eaton (unpublished) have developed an HPLC assay for measuring all metabolites of methyl parathion and showed that dealkylation to des-methylparathion occurs readily in the cytosolic fraction of rat liver in the presence of GSH; the extent of activity in peripheral blood lymphocytes has not yet been determined.

Other substrates are of interest, as well. Seidegard et al reported that 65% of 66 lung cancer patients had the low GST phenotype with trans-stilbene oxide, compared with 41% of control subjects with similar smoking histories (22). They postulated that lack of adequate Phase II GSH conjugation might have enhanced susceptibility of smokers to lung cancer. There are two problems in this conclusion: first, the prevalence of the low activity GST phenotype in unaffected smokers (41%) was far below the 54% previously reported (21); and, second, trans-stilbene oxide is not a very likely substrate to correlate with exposures from cigarette smoking. Eaton and Richards (unpublished results) have now shown that human mononuclear leukocytes at pH 7.4 have readily detectable activity against BP-4,5-oxide and that the activity tends to distribute with the trimodality described with trans-stilbene oxide as substrate. This finding opens the way to considerable additional research using the peripheral cells to investigate metabolism of key xenobiotic agents.

INCORPORATION OF DATA ON GENETIC VARIATION INTO RISK ASSESSMENT FOR PESTICIDES

The U.S. Environmental Protection Agency and its counterparts in many European countries now routinely review epidemiologic, animal bioassay, and in vitro test results in the approval of applications for registration of new pesticides and in reviews of pesticides in use. When the 95 percent upper confidence estimate of lifetime carcinogenic risk, based upon a linear or linearized multistage extrapolation to low-dose estimates for population exposures, exceeds one additional cancer case per 100,000 exposed persons, the Agency is predisposed to take action. Our laws require that the economic benefits of pesticides be considered when risks are reviewed. It is a useful guideline to note that EPA regularly bans or takes other strong action against chemicals with calculated upper-bound risks exceeding 1/1000 and seldom does so when the estimated risk is less than 1/100,000. Within the 1/1,000 to 1/100,000 range, other factors are highly influential.

Marked prominently in our framework (Figure 1), variation in susceptibility may be due to age, sex, nutrition, co-existing exposures, pre-existing diseases, medications, vitamins, use of protective gear, or inherited factors affecting metabolism of the chemical or target-site response to the chemical or its active metabolites. The Clean Air Act and the Occupational Safety & Health Act in the U.S. explicitly require that standards be set so as to protect the most susceptible subgroups or individuals, respectively (23). In general, however, relatively little is known about such differences in susceptibility to actual exposures.

An excellent prospect for investigating the role of such potential genetic differences in susceptibility to specific chemicals would be worker exposures to parathion. The metabolism of parathion is complicated, but the main steps are the P450 oxidation of parathion or paraoxon and the hydrolysis of paraoxon to p-nitro-phenol and diethylphosphate (Figure 2). Ideally, we should have in hand practical assays for genetic variability in the relevant P450 isozymes and in the glutathione S-transferases (less important for parathion than for methylparathion). Good assays for paraoxonase exist already. Adequate methods for measurement of ambient and personal exposures exist. Subclinical toxicity can be assessed through the primary biological endpoint, inhibition of acetyl-cholinesterase, in red blood cells. Finally, clinical neurological status can be evaluated.

We postulate that individuals with similar exposures would be more susceptible to AChE inhibition and development of clinical neurotoxicity if their paraoxonase phenotype were low-activity, rather than high or intermediate activity. Obviously, if toxicity resulted not from ordinary exposures, but only from accidental major spills, then the effect of the genetic difference in inactivation of the toxic metabolite might be less important or negligible. But even with spills workers are observed to vary markedly in their extent of symptoms and the time course of their recovery. On the other extreme, if experienced applicators seldom had accidental spills, but always had a low level of exposure, the metabolic variation might be quite important. Assay of the P450 and perhaps also the glutathione S-transferase steps would be important primarily in explaining other host variation. Paraoxon is the toxic agent and its formation and especially its hydrolysis would be the main matter of interest.

Identification of workers at higher risk due to low activity of paraoxonase could be useful in worker-education programs about safe use of pesticides and could be incorporated into decision rules about the frequency of monitoring of AChE levels and the timing of return to work during the recovery phase after any acute exposure. It would be neces-sary to estimate the proportion of the risk of adverse effects that might be attributed to genetic variation in susceptibility, compared with variation in exposure and variation in non-genetic host factors. Unless the attributable risk were impressive, perhaps 25 percent of the overall variation in risk, and the toxicity actually were observed occasionally, there might be little interest in embarking on a path toward incorporat-ing such a genetic test on a screening basis for new or current employ-ees. Any such screening effort must have the full informed consent of the workers and their representatives; management and those tested must know in advance how the information would be used and who would have access to it.

For the clinical occupational physician, it would be useful to have such tests for investigation of the patient who asks "Why me, Doc?" upon learning that his symptoms may be due to exposures at the worksite. Also such tests might be very useful for epidemiologic investigations of possible effects of low-dose chronic exposures, including many unexplained neurologic syndromes. In the example noted above Barbeau et al (4) postulated that environmental chemical neurotoxins might be involved in the development of Parkinson's disease, especially cases with onset before age 55, and that genetic differences in the P450 oxidation system might indicate differential susceptibility. Costa in this volume has reported experiments in animals in which behavioral and cognitive abnormalities developed from repeated doses not associated with acute toxicity. During informal discussions at this meeting, colleagues from Denmark, Italy, and China have reinforced my impression that there may be many workers with chronically depressed AChE levels potentially at risk for later cognitive deficits.

Smith has noted (24) that drugs which are substrates for the debrisoquine-related P450 are relatively strongly basic compounds. Outstanding in their affinity for this metabolizing enzyme system are natural plant alkaloids, which include sparteine, quinidine, ajmaline, lobeline, papaverine, and yohimbine. Perhaps the debrisoquine gene locus evolved to deal metabolically with toxic food constituents, particularly plant alkaloids. Also Smith noted that the highly toxic tetrodotoxins of various marine and amphibian animals resemble debrisoquine as strongly basic guanidine derivatives; their toxicity is due to interference with Na^+ channels in nerve membranes. These speculative points reinforce the possibility that the P450 system and especially the debrisoquine-related P450 may be a key determinant of susceptibility to environmental chemicals that affect the nervous system.

With the dramatic advances in knowledge of genetics and in genetic techniques, we can anticipate that for pesticides and for other chemicals we will increasingly become able to ascertain differences in metabolism and in susceptibility to target organ actions. In time we should be able to incorporate such information into risk assessment and risk reduction efforts.

REFERENCES

1. Omenn GS, Motulsky AG. Eco-genetics: genetic variation in susceptibility to environmental agents. IN Genetic Issues in Public
Health and Medicine. BH Cohen, AM Lilienfeld, PC Huang (eds).
Springfield, Ill.: CC Thomas, 1978, pp. 83-111.
2. Omenn GS and Gelboin H (eds). Genetic Variability in Responses to
Chemical Exposures, Banbury Report 16. Cold Spring Harbor
Laboratory, New York, 1984.
3. Neal RA. Studies of the enzymic mechanism of the metabolism of
diethyl 4-nitrophenyl phosphorothionate (parathion) by rat liver
microsomes. Biochem J, 105:289-297, 1967.
4. Barbeau A, Rom M, Paris S, Cloutier T, Plasse L, Poirier J. Eco-
genetics of Parkinson's disease: 4-hydroxylation of debrisoquine.
Lancet ii:1213-1215, 1985.
5. Syvalahti EKG, Lindberg R, Kallio J, DeVocht M. Inhibitory effects
of neuroleptics on debrisoquine oxidation in man. Brit J Clin
Pharmacol, 22:89-92, 1986.
6. Distlerath LM, Larrey D, Guengerich FP. Genetic polymorphism of
debrisoquine-4-hydroxylation: identification of the defect at the
level of a specific cytochrome P-450 in a rat model. IN Genetic
Variability in Responses to Chemical Exposures, Omenn GS and
Gelboin H (eds). Cold Spring Harbor Laboratory, New York, 1984,
pp. 85-95.
7. Guengerich FP, Martin MV, Beaune PH, Kremers P, Wolff T, Waxman DJ.
Characterization of rat and human liver microsomal cytochrome
P-450 forms involved in nifedipine oxidation, a prototype for
genetic polymorphism in oxidative drug metabolism. J Biol Chem,
261:5051-5060, 1986.
8. Neal RA. A comparison of the in vitro metabolism of parathion in
the lung and liver of the rabbit. Toxicol App Pharmacol, 23:123-
130, 1972.
9. Nebert DW. Genetic differences in susceptibility to chemically-
induced myelotoxicity and leukemia. Environ Health Perspectives,
39:11, 1981.
10. Kouri RE, McLemore T, Jaiswal AK, Nebert DW. Current cellular
assays for measuring clinical drug metabolizing capacity—impact
of new molecular biologic techniques. IN Ethnic Differences in
Reactions to Drugs and Xenobiotics, Kalow W, Goedde HW and Agarwal
DP (eds). Alan R. Liss, New York, 1986, pp. 453-469.
11. Koop DR, Crump BL, Nordblom GD, Coon MJ. Immunochemical evidence
for induction of the alcohol-oxidizing cytochrome P-450 of rabbit
liver microsomes by diverse agents: ethanol, imidazole, trichloroethylene, acetone, pyrazole, and isoniazid. Proc Natl Acad Sci
U.S., 82:40654069, 1985.
12. LaDu BN, Adkins S, Bayoumi RA-L. Analysis of the serum paraoxonase/
arylesterase polymorphism in some Sudanese families. IN Ethnic
Differences in Reactions to Drugs and Xenobiotics, Kalow W, Goedde
HW and Agarwal DP (eds). Alan R. Liss, New York, 1986, pp. 87-98.
13. Mueller RF, Hornung S, Furlong CE, Anderson J, Giblett ER, Motulsky
AG. Plasma paraoxonase polymorphism: a new enzyme assay, population, family, biochemical, and linkage studies. Amer J Human
Genet, 35:393-408.
14. Ortigoza-Ferado J, Richter R, Furlong C, Motulsky AG. Biochemical
genetics of paraoxonase. IN Genetic Variability in Responses to

Chemical Exposure, Omenn GS and Gelboin HV (eds). Cold Spring Harbor Laboratory, New York, 1984, pp. 177–188.

15. Costa LG, Richter RJ, Murphy SD, Omenn GS, Motulsky AG, Furlong CE. Species differences in serum paraoxonase correlate with sensitivity to paraoxon toxicity. IN Toxicology of Pesticides: Experimental, Clinical, and Regulatory Perspectives (this volume).

16. Arias IM and Jakoby WB (eds). Glutathione: Metabolism and Function. Raven Press, New York, 1976.

17. Fukami J-I. Metabolism of several insecticides by glutathion-S-transferase. Pharmac Ther, 10:473–514, 1980.

18. Motoyama N, Dauterman WC. Glutathione S-transferases: their role in the metabolism of organophosphorus insecticides. Review in Biochemical Toxicology, Hodgson E, Bend JR, and Philpot RM (eds). Elsevier/North Holland, New York, 1980, pp. 49–69.

19. Board PG. Biochemical genetics of glutathione-S-transferase in man. Amer J Hum Genet, 33:36–43, 1981.

20. Strange RC, Faulder CG, Davis BA, Hume R, Brown JAH, Cotton W, Hopkinson DA. The human glutathione S-transferases: studies on the tissue distribution and genetic variation of the GST1, GST2 and GST3 isozymes. Ann Hum Genet, 48:11–20, 1984.

21. Seidegard J, Pero RW. The hereditary transmission of high glutathione transferase activity towards trans-stilbene oxide in human mononuclear leukocytes. Hum Genet, 69:66–68, 1985.

22. Seidegard J, Pero RW, Miller DG, Beattie EJ. A glutathione transferase in human leukocytes as a marker for the susceptibility to lung cancer. Carcinogenesis, 7:751–753, 1986.

23. Omenn GS. Genetics and epidemiology: medical interventions and public policy. Soc Biol, 26:117–125, 1979.

24. Smith RL. Introduction to special issue on polymorphisms of drug oxidation. Xenobiotica, 16:361–365, 1986.

THE REPRODUCTIVE TOXICOLOGY OF PESTICIDES

Neil Chernoff
United States Environmental Protection Agency
MD-2
Research Triangle Park, NC   27711   USA

The successful reproduction of any species depends upon the accurate
and orderly integration of genetic, biochemical, physiological, developmental, and functional processes. Any significant interference with the normal functioning of the organism may manifest itself as an adverse effect
on reproduction. While this statement is an extremely obvious one, it is
important to bear in mind given the design and endpoints of some of the
standard protocols used to assess reproductive toxicity. If we strictly
define the processes which are essential for mammalian reproduction we may
list the following components: genetic integrity; gametogenesis; transportation of the gametes; fertilization; formation of the pre-implantation
embryo; maternal hormonal, physiological, and structural alterations necessary for in utero development; implantation of the embryo; placentation;
embryogenesis; fetal growth; parturition and the changes necessary for the
fetus to assume an independent existence; maternal/neonatal relationship;
and postnatal growth and maturation. In a practical (or regulatory) sense,
reproductive toxicity is divided into two areas: the potential of agents
to adversely affect either structural or functional development (teratogenicity), or any of the other processes listed above (reproductive toxicity).

Prior to any review of the relevant literature, a brief dission of the
standard protocols utilized to determine reproductive toxicity is desirable.
Generally, reproductive toxicity is evaluated by variations on one of three
types of bioassays dealing respectively with 1. general fertility and
reproduction in which animals are exposed to the agent and the viability and
the fertility of the offspring are measured (multigeneration study); 2. a
test in which the agent is administered to the pregnant animal and fetal

NATO ASI Series, Vol. H13
Toxicology of Pesticides: Experimental, Clinical
and Regulatory Aspects. Edited by L.G. Costa et al.
© Springer-Verlag Berlin Heidelberg 1987

anomalies are recorded; (teratology or developmental toxicology study); and 3. a study in which animals are exposed to the test compound late in gestation and through the lactational period (perinatal study).

The multigeneration study consists of multiple dose groups and concurrent controls which receive the test substance through a premating period that covers both the spermatogenic and estrus cycles of the animals (generally the rat, although other rodent species have been used). The precise design of the study depends upon the country it is done in and varies from a fertility study in which the compound is administered only through the early gestational period (Japan) to a multigeneration study (as is done in the U.S.A. and Great Britain). There has been considerable debate concerning both interim sacrifices of pregnant females during gestation, and the need for more than one or two generations (Clegg, 1979). Generally a minimum of two litters for each generation is recommended and in some protocols histopathology is done on offspring of the last generation. A detailed listing of the designs of the different protocols currently in use can be found in the ECETOC (1983) document. There has been considerable discussion of the benefits of the different experimental designs used in multigeneration studies (Palmer, 1981; IPCS, 1984; Christian, 1986) and there has been, in recent years, some movement towards concurrence of the basic design - one which omits the third generation and the interim sacrifice of females for examination of their litters. I would like to emphasize the following point: the multigeneration study as it is presently constituted does not identify agents that adversely affect the reproductive system in a strictly defined sense. These studies are perhaps more properly thought of as tests of the bioactivity of agents across generations. Sick animals may lose their ability to breed by innumerable factors not directly dealing with any of those listed above as part of the continuum needed for successful reproduction. Similarly, a positive finding in this study often revolves about the loss of pups but unless we are aware of damage to the reproductive system itself all that we know is that the agent being tested has caused some severe adverse effects that have manifested themselves in death of the offspring.

There have been a number of symposia and articles in recent years which have focused upon the need for test protocols that will identify specific reproductive system components that may be the targets of xenobiotics. Among the parameters that have been suggested to accomplish this aim are the identification of estrogen and androgen agonistic-antagonistic activity; dysfunctions in the hypothalamic-pituatary-gonadal axis; alterations in reproductive behavior; ovarian toxicity; inhibition of steroidogenesis, alterations in seminal fluid composition; and various measures on critical parameters of sperm including numbers, morphology, motility, and ability to fertilize. The identification and validation of suitable tests to evaluate these aspects of the reproductive cycle will enable both the identification of compounds capable of selectively affecting reproductive parameters and the specific sites of injury (E.P.A. 1980; Dixon and Hall, 1982).

The protocol utilized for the assessment of an agent's potential to affect embryo/fetal development involves the exposure of pregnant animals to the test compound throughout the period of major organogenesis, gathering the fetuses shortly before parturition, and the subsequent examination of the fetuses for both soft tissue and skeletal anomalies. Generally two species are called for, and multiple dose levels are used - one of which is high enough to induce some form of overt maternal toxicity. Examples of such protocols may be found in a number of sources including the Canadian (1977), OECD (1981) and the EPA (1982). There are some basic limitations of the standard protocol which should be noted. The fetal examination is entirely morphological and therefore may both miss subtle or small defects in organs. and fail to identify either critical organ toxicity in the fetus or biochemical defects that may only manifest themselves as postnatal functional deficits. The utilization of dose levels which induce maternal toxicity raises the question of the potential of such toxicity to affect the embryo/fetal response in a general fashion. Related to this last point is the question of the potential role that compound-induced maternal nonspecific stress may play in the production of fetal anomalies.

Recent work has indicated that the concerns raised above may indeed be significant factors in our assessment of teratology data. Kavlock et al. (1982) have shown that compounds are capable of affecting critical biochemical parameters of selected fetal organs at dose levels below those required to elicit gross morphological alterations. That there are numerous chemicals which produce in utero effects manifesting themselves as postnatal functional deficits is becoming clear with the work of many researchers who have investigated the functions of a variety of organ systems during the postnatal period. Abnormal functional development resulting from in utero exposure has been iden-

tified in the cardiac, renal, pulmonary, central nervous, and other systems
(see Kimmel and Buelke-Sam, 1981; and Kavlock and Grabowski, 1983). The re-
lationship(s) of maternal and embryo/fetal toxicity has been examined by
both literature review and experimentation. Khera has found an association
between specific types of anomalies and the presence of overt maternal toxicity in standard teratology tests published in the open literature (Khera,
1985) although the work of Kavlock et al. (1985) on acute maternal toxicity
and fetal development indicated only chemical specific effects with the exception of an increased incidence of supernumerary ribs raising the question
of effects of general maternal stress on the production of birth defects.
Adverse embryo/fetal effects have been noted in litters of dams subjected to
various forms of stress including noise (Nawrot et al., 1981) and restraint
(Barlow et al., 1975;). The results in these studies have varied with the
form of stress and the species used. Recent work has shown that restraint
stress produces supernumerary ribs encephaloceles, and fused ribs in mice
but is without effects in rats (Beyer and Chernoff, 1986). The degree of
stress as measured by appropriate biochemical parameters has not been measured in any studies done to date so no definitive statements can be made
concerning possible comparisons between the stresses used above and those
produced by toxic levels of xenobiotics. The above work points out that we
must interpret the standard teratology protocol carefully and in many cases,
especially where we have overt maternal toxicity and/or no postnatal information, additional studies are needed as predicated by the data generated.

The third type of test, a perinatal study, involves dosing during late
gestation through the lactational period. As Palmer (1981) points out, such
studies may serve to identify those compounds whose effects take place during this period rather than seeing such effects as part of the more expensive
and laborious multigeneration study. Such studies are not universally called
for, and it might be useful to attempt to form a consensus on their need and
design. It has been suggested that some form of postnatal study be a part of
every standard teratology test thus enabling investigators to identify effects that are not readily apparent with fetal examination as well as to obtain information on the significance of a variety of fetal effects that are
difficult to interpret in a regulatory setting (e.g. supernumerary ribs,
enlarged renal pelvis, and reduced fetal weight). The above discussion
should serve as an introduction to a general review of the potential of
pesticides to adversely affect the reproductive cycle of mammals. It is

beyond the scope of this paper to detail all of the numerous studies that have been published on this subject, and my intent is to examine a selected body of data to illustrate the variety of pesticides found to have adverse reproductive effects, as well as the types of tests which have been used to show such toxicity.

There have been numerous multigeneration studies performed on pesticides, the majority done to comply with regulatory needs. Examples of this type of study include those done by Collins et al. (1971) on the insecticide Carbaryl. These experiments followed standard U.S.A. Food and Drug Administration protocols for three generation reproductive toxicity studies. These scientists found a significant weight reduction in treated pups at all dose levels. At the two highest dose levels the authors noted impaired fertility, and dose-related decreases in litter size, number of liveborn, and postnatal viability. Finally, there was a significant growth depression seen in maternal animals at the highest dose level. Murray et al. (1979) studied the reproduction of rats given 2,3,7,8-tetrachlorodibenzo-p-dioxin (TCDD) which may be a contaminent of compounds made from 2,4,5-trichlorophenol including the herbicide 2,4,5-trichlorophenoxyacetic acid. They found that administration of this chemical to rats across three generations produced adverse reproductive effects as seen in reduced fertility, litter size, and postnatal growth. These effects are similar to those noted in an earlier study by Khera and Ruddick (1972) in which TCDD was administered to pregnant rats during gestation days 6-15 and indicate that TCDD causes fetal death and perinatal growth retardation. The studies reviewed above indicate their utility in the determination of the biological activity of the compounds studied. They give clear evidence that these compounds interfere with normal biological functioning at some dose levels, but it should also be noted that there is no evidence of selective effects on any reproductive parameter. Whether the effects noted by both Collins et al. (1971) and Murray et al. (1979) are due to general toxicity or selective effects on components of the reproductive system remains unknown. The multigeneration study in concert with other tests of chemicals' biological activities allows us to make rational assessments on their potential health effects, and may serve to direct the course of further tests. For information on the reproductive system itself, other studies especially designed to evaluate specific components of this system may be necessary.

The critical role that estrogens play in the reproductive cycle is well known and pesticides that have estrogenic properties have been demonstrated to affect reproduction. Those pesticides have been shown to directly affect the female reproductive system include DDT, chlordecone, methoxychlor, and mirex. Heinricks et al. (1971) administered o,p'-DDT to pre-pubertal rats and showed that treated females exhibited precocious puberty, increased uterine weight, and persistent vaginal estrus. Chlordecone induces defeminization, and persistent vaginal estrus when administered to rats during the period of behavioral differentiation (Gellert, 1978). Similarly, Gray (1982) showed that the administration of chlordecone to neonatal hamsters results in masculinization of females resulting in persistent aberrant sexual behavior. Dietary exposure of methoxychlor to rats before mating resulted in early vaginal opening and reduced reproduction in female offspring (Harris et al., 1974). The relatively few pesticides which have been shown to directly affect the male reproductive system are generally chlorinated hydrocarbons with estrogenic properties. Whether the relative lack of compounds shown to adversely affect females reflects the relative ease with which the male reproductive system may be studied, or the relative insensitivity of the female system is not known at this time.

Numerous pesticides have been shown to significantly alter male reproductive parameters in laboratory studies. As would be expected, compounds which exhibit estrogenic properties may affect both female and male reproductive parameters. DDT exposure in male dogs results in decreased reproductive capacity (Deichmann et al., 1971), reduced uptake of testosterone in the prostate in mice, and o,p'-DDT has been shown to affect androgen metabolism in man (Hellman et al., 1973). The adverse effects of chlordecone on the male reproductive systems in laboratory animals as well as workers exposed to large quantities of this pesticide, have been well documented. Exposure to this pesticide in adult rats resulted in transient decreases in both viability and motility of epididymal spermatozoa along with decreased sperm reserves in the cauda epididymus (Linder et al., 1983). Male workers who were exposed to large amounts of chlordecone during the manufacture, exhibited adverse central nervous system, course of its hepatic, and reproductive effects. The effects on sperm were similar to those later shown in laboratory studies and included decreased counts and motility as well as an elevated incidence of abnormal morphology (Taylor et al., 1978). The cyclodiene pesticides dieldrin (Schein and Thomas, 1975) and chlordane (Levin et al., 1969), and the herbicide 2,4,5-trichlorophenoxyacetic acid (Lloyd et al., 1973) have all been shown to alter testosterone kinetics but the effects (if any) on reproductive function remain unknown. The organophosphate dichlorvos has been tested in rats (Krause and Homola, 1974) and histopathological examination revealed degeneration of the seminiferous tubules accompanied by significant reductions of spermatozoa and spermatids. The bipyridal herbicides diquat and paraquat were found to induce significant adverse effects on the development of sperm when administered to mice (Pasi et al., 1974). Prenatal treatment with the diphenyl ether herbicide nitrofen early in gestation results in anomalies in the male reproductive tract while treatment later in gestation produces behavioral alterations and reduced fertility in male offspring (Gray et al., 1985). The dinitrophenol herbicide dinoseb has been shown to induce severe morphological alterations in sperm coupled with significant oligospermia when the compound was administered to the rat in the diet (Linder et al., 1982). The fumigant ethylene dibromide produced oligospermia and degeneration of existing spermatozoa in bull calves fed the pesticide for a year following birth (Amir and Volcani., 1967). Another fumigant, 1,2-dibromo-3-chloropropane (DBCP) has been shown to produce adverse male reproductive effects in both laboratory animals and in exposed humans. Testicular atrophy has been documented after inhalation exposure in rats (Rao et al., 1983). This pesticide also produced reduced fertility in occupationally exposed workers, and testicular biopsies in affected individuals indicated significant oligospermia (Whorton et al. 1979). The fungicide benomyl has been shown to exhibit age-related male reproductive toxicity. The oral administration of this pesticide in the pubertal or postpubertal rat results in decreased testicular weight, and decreased sperm in the epididymus and vas deferens. Similar administration in the prepubertal animal did not induce these alterations (Carter et al., 1984).

A recent review by Schardein (1985) listed approximately 200 pesticides or their degradation products as having been tested for their teratogenic potential. Of these, some 90 were indicated as having one or more studies which have resulted in positive teratogenic effects under the experimental regimens used. There are to date, however, no well documented associations between the human exposure to pesticides and the subsequent occurrence of birth defects. This apparent disparity between positive effects noted in animals studies and the lack of known human effects, is similar to the general situation with the teratogenic potential of all types of agents, and it should be noted, for other aspects of reproductive toxicology. Shepard (1986) estimates that some 900 agents listed in his catalogue have been shown to produce birth defects in the laboratory setting while only between 20 and 30 agents are known to be teratogenic in humans. These differences in response(s) may have a number of basic causes. Interspecies differences may be the result of differences in maternal metabolism and/or sensitivity, embryo/fetal developmental patterns, and placentation. The major contributor to the apparent interspecies difference is most probably the fact that virtually all standard teratology bioassay protocols used to register pesticides utilize a high dose which induces some form of overt maternal toxicity so as to maximize the chances of seeing an effect with a small number of animals. These dose levels are invariably far greater than that which would be encountered by humans during appropriate environmental exposure. When the results of such studies are considered in a "margin of safety" framework, they provide for a scientifically reasonable approach to the protection of the human population. It should be noted that for an agent to be firmly identified as a human teratogen, a number of practical criteria may have to be met. The agent has to either produce a large increase in defects which would be noted by alert health personnel as in the case of rubella (Gregg, 1941), or a type of defect of such an unusual nature that it would again be noted by health personnel such as was seen with thalidomide (McBride, 1961). The epidemiological identification of a specific pesticide which produced a relatively small increase in a variety of common birth defects would be extraordinarily difficult to achieve given the complexity of the human environment.

The following review of the teratogenicity of pesticides will, of necessity, will concentrate on commonly used pesticides, and those which illustrate some of the problems alluded to above. The carbamate insecticide carbaryl has been extensively studied by many workers in a range of species including the rat, mouse, guinea pig, rabbit, and dog (see Cranmer, 1986). Adverse embryo/fetal effects were seen in a number of these species including fetal death in the rat and rabbit, skeletal defects in the guinea pig, and a broad spectrum of severe defects in the dog including agenesis of the external genitalia, pubis and ischium, and various viscera. When these data are examined a pattern is seen which indicates that these defects were only noted at dose levels at which overt maternal toxicity was present. The maternal toxicity included lethality, decreased weight gain, clinically observable cholinesterase inhibition, and dystocia. These data indicate that the developing organism is not especially sensitive to carbaryl, but rather, that the doses required to elicit developmental toxicity are similar to those resulting in maternal toxicity and are far in excess of any that might be encountered in the environment. It has generally been concluded therefore, that although carbaryl is known to be fetotoxic/teratogenic in a number of species it does not present an unreasonable risk to pregnant humans in the general population under approved patterns of use.

The cyclodiene insecticide endrin produced neural defects in the absence of maternal toxicity after single administration on day 8 in the hamster (Ottolenghi et al., 1974). When the compound was administered during the standard period, days 4-15 (Chernoff et al., 1979), the neural defects noted in the former study were only noted at dose levels that were maternally toxic. Kavlock et al., (1981) studied the teratogenic potential of endrin in mice and their studies indicated that in this species also, fetal toxicity was present only at maternally toxic dose levels. Here too, as in the case of carbaryl, embryo/fetal and adult toxic dose levels are similar when standard periods of prenatal administration are used.

The herbicides 2,4,5-trichlorophenoxyacetic acid (2,4,5-T) and 2,4-dichlorophenoxy acetic acid (2,4-D) are widely used and have been extensively tested for their teratogenic potential. Early work which found significant increases in fetal toxicity in rats and mice (Courtney et al., 1970) may have been confounded by the presence of significant levels of 2,3,7,8-tetrachloro-di-benzodioxin (TCDD). Studies with the purified herbicide have shown that at dose levels approaching maternal toxicity, 2,4,5-T is fetotoxic in the rat (Sparschu et al., 1971). In the mouse, Neubert and Dillmann (1972) demonstrated that this compound produced reductions in fetal weight and cleft palate. Nelson et al. (1979) reviewed the epidemiological data on this chemical and reported on such a study that they had performed. They concluded that there was no demonstrable relationship(s) between the incidence of birth defects and exposure to 2,4,5-T. The closely related herbicide, 2,4-D has been found to produce fetotoxicity in rats including wavy ribs, edema, and delayed ossification (Schwetz et al., 1971).

The fungicide benomyl is teratogenic in both the mouse and rat, producing a wide spectrum of malformations when it was administered by oral gavage. The administration of benomyl via dietary exposure was an order of magnitude less effective in producing fetal effects when maternal intakes on a mg/kg/day basis are compared (Kavlock et al., 1982). This study illustrates the potential importance of the route of administration in the determination of fetal outcome after exposure to an agent. Captan has been widely studied in large part because of its structural relationship to the potent human teratogen, thalidomide. Results to date have indicated that this chemical is not teratogenic in rabbits (Fabro et al., 1966) or rats (Kennedy et al., 1968). Maneb is a dithiocarbamate fungicide that has been found to be teratogenic in rats (inducing hydrocephaly) but not in the mouse or hamster. These results parallel the effects seen after exposure to ethylene thiourea, a primary degradation product, in these species (Chernoff et al., 1979).

There has been, in recent years, an increasing tendency of teratologists to examine postnatal animals for signs of prenatally induced injury. Such studies have indicated that postnatal evaluations often yield information not readily obtained during the course of standard fetal examinations. The prenatal administration of the herbicide nitrofen results in ablation of the Harderian gland, an effect which cannot be identified during fetal examination (Gray et al., 1983). In the case of benomyl (discussed above) it is of note that the most sensitive indicator of perinatal exposure was a permanent reduction in testes and accessory gland weights Kavlock et al., 1982). Examination of postnatal animals treated with the insecticide mirex has revealed that this compound induces cataract formation with in utero (Rogers and Grabowski, 1983) or neonatal exposure through the maternal milk (Scotti et al., 1981), an effect not readily noted during fetal examination.

In this paper I have attempted to give an overview of the reproductive toxicity of pesticides. I have tried to illustrate the wide variety of effects that have been noted in the laboratory setting, as well as in those few instances where deleterious human effects have been conclusively shown. I have also attempted to utilize compounds and/or studies which highlight some of the common problems facing regulatory personnel in their attempts to extrapolate laboratory studies to human populations - questions of relevance of effects, appropriate routes of administration and dose levels, interspecies differences, and limitations inherent in the commonly used protocols.

LITERATURE CITED

Amir D, Volcani R (1967) Effect of dietary ethylene dibromide (EDB) on the testes of bills. Fertil Steril 18:144-148

Barlow SM, McElhatton PR, Sullivan FM (1975) The relation between maternal restraint and food deprivation, plasma corticosterone, and induction of cleft palate in the offspring of mice. Teratol 12:97-104

Beyer PE, Chernoff N (1986) The induction of supernumerary ribs in rodents: Role of maternal stress. Terato, Carcino and Mutagen (in press)

Canadian Ministry of Health and Welfare, Health Protection Branch (1977) The testing of chemicals for teratogenicity, mutagenicity and carcinogenicity

Carter SD, Hein JF, Rehnberg GL, Laskey JW (1984) Effect of benomyl on the reproductive development of male rats. J Toxicol Environ Hlth 13:53-68

Chernoff N, Kavlock RJ, Hanisch RC, Whitehouse DA, Gray JA, Gray LE Jr, Sovocool GW (1979) Perinatal toxicity of endrin in rodents. I. Fetotoxic effects of prenatal exposure in hamsters. Toxicol 13:155-165

Chernoff N, Kavlock RJ, Rogers EH, Carver BD, Murray S (1979) Perinatal toxicity of maneb, ethylene thiourea, and ethylene-bisisothiocyanate sulfide in rodents. J Toxicol Environ Hlth 5:821-834

Christian MS (1986) A critical review of multigeneration studies. J Amer Coll Toxicol 5(2):161-180

Clegg DJ (1979) Animal reproduction and carcinogenicity studies in relation to human safety evaluation. In: Deichmann WB (ed) Toxicology and occupational medicine. WHO

Collins TFX, Hansen WH, Keeler HV (1971) The effect of carbaryl (Sevin) on reproduction of the rat and the gerbil. Toxicol Appl Pharmacol 19:202-216

Courtney KD, Gaylor DW, Hogan MD, Falk HL, Bates RR and Mitchell I (1970) Teratogenic evaluation of 2,4,5-T. Science 168:864-866 Weil CS,

Cranmer MF (1986) CARBARYL A toxicological review and risk analysis. Neurotox 7:1-79

Deichmann WB, Macdonald WE, Beasley AG and Cubit D (1971) Subnormal reproduction in beagle dogs induced by DDT and aldrin. Ind Med Surg 40:10-20

Dixon RL, Hall JL (1982) Reproductive toxicology. In: A. Wallace Hayes (ed) Principles and Methods of Toxicology, Raven Press New York, NY

European Chemical Industry Ecology and Toxicology Centre (1983) Identification and assessment of the effects of chemicals on reproduction and development (Reproductive Toxicology) Monograph #5

Fabro S, Smith RL, Williams RT (1966) Embryotoxic activity of some pesticides and drugs related to phthalimide. Food Cosmet Toxicol 3:587-590

Gellert RJ (1978) Kepone, Mirex, Dieldrin and Aldrin: Estrogenic activity and the induction of persistent vaginal estrus and anovulation in rats following neonatal treatment. Environ Res 16:131-138

Gray LE Jr (1982) Neonatal chlordecone exposure alters behavioral sex differentiation in female hamsters. Neurotoxicol 3:(2)67-80

Gray LE, Kavlock RJ, Chernoff N, Ostby J, Ferrell J (1983) Postnatal developmental alteration following prenatal exposure to the herbicide 2,4-dichlorophenyl-p-nitrophenyl ether: a dose response evaluation in the mouse. Toxicol Appl Pharmacol 67:1-14

Gray LE Jr, Ferrell J, Ostby J (1985) Prenatal exposure to nitrofen causes anomolous development of para- and mesonephric duct derivatives in the hamster. Toxicologist 5:(1)183

Gregg NM (1941) Congenital cataract following German measles in the mother. Trans Opthalmol Soc Aust 3:35-46.

Harris SJ. Cecil HC, Bitman J (1974) Effect of several dietary levels of technical methoxychlor on reproduction in rats. J Agric Food Chem 22:969-973

Heinricks WJ, Gellert RJ, Bakke JL, Lawrence NL (1971) DDT administered to neonatal rats induces persistent estrus syndrome. Science 173:642-643

Hellman L, Bradlow HL, Zumoff B (1973) Decreased conversion of androgens to normal 17-ketosteroid metabolites as a result of treatment of o,p'-DDD. J Clin Endocrinol Metab 36:801-803

IPCS International Programme on Chemical Safety, Environmental Health Criteria 30 - Principles for evaluating health risks to progeny associated with exposure to chemicals during pregnancy, WHO (1984)

Kavlock RJ, Chernoff N, Hanisch RC, Gray JA, Rogers E, Gray LE Jr (1981) Perinatal toxicity of endrin in rodents. II. Fetotoxic effects of prenatal exposure in rats and mice. Toxicol 21:141-150

Kavlock RJ, Chernoff N, Gray LE Jr, Gray JA, Whitehouse D (1982) Teratogenic effects of benomyl in the Wistar rat and CD-1 mouse, with emphasis on the route of administration. Toxicol Appl Pharmacol 62:44-54

Kavlock RJ, Grabowski CT (eds) (1983) Abnormal functional development of the heart, lungs, and kidneys. Alan R Liss Inc, New York

Kavlock RJ, Chernoff N, Rogers E (1985) The effect of acute maternal toxicity on fetal development in the mouse. Terato, Carcino and Mutagen 5:3-15

Kennedy GL, Fancher OE, Calandra JC (1968) An investigation of the teratogenic potential of Captan, Folpet and Difolatan. Toxicol Appl Pharmacol 13:420-430

Khera KS, Ruddick JA (1972) Polychlorodibenzo-p-dioxins: Perinatal effects and the dominant lethal test in Wistar rats. Chlorodioxins-Origin and Fate (EH Blair, ed) Advances in Chemistry Series 120. Amer Chem Soc, Washington, DC

Khera KS (1985) Maternal toxicity - A possible etiological factor in embryo-fetal deaths and fetal malformations of rodentrabbit species. Teratol 31:129-153

Kimmel CA, Buelke-Sam, J (eds) (1981) Developmental toxicology. Raven Press, New York

Krause W, Homola S (1974) Alterations of the seminiferous epithelium and the Leydig cells of the rat testis after the application of dichlorvos (DDVP). Bull Environ Contam Toxicol 11:429-433

Levin W, Welch RM, Conney AH (1969) Inhibitory effect of phenobarbital or chlordane pretreatment on the androgeninduced increase in seminal vesicle weight in the rat. Steroids 13:155

Linder RE, Scotti TM, Svendsgaard DJ, McElroy WK, Curley A (1982) Testicular effects of dinoseb in the rat. Arch Environ Contam Toxicol 11:475-485

Linder RE, Scotti TM, McElroy WK, Laskey JW (1983) Spermotoxicity and tissue accumulation of chlordecone (Kepone) in male rats. J Toxicol Environ Hlth 12:183-192

Lloyd JW, Thomas JA, Mawhinney MC (1973) 2,4,5-T and the metabolism of testosterone-1,2-^{3}H by mouse prostate glands. Arch Environ Hlth 26:217

McBride, WG (1961) Thalidomide and congenital abnormalities. Lancet 2:1358

Murray FJ, Smith FA, Nitschke KD, Humiston CG, Kociba RJ, Schwetz BA (1979) Three-generation reproduction study of rats given 2,3,7,8-tetrachlorodibenzo-p-dioxin (TCDD) in the diet. Toxicol Appl Pharmacol 50:241-252

Nawrot PS, Cook RO, Hamm, CW (1981) Embryotoxicity of broad-band high-frequency noise in the CD-1 mouse. J Toxicol Environ Hlt 8:151-157

Nelson CJ, Holson JF, Green HG, Gaylor DW (1979) Retrospective study of the relationship between agricultural use of 2,4,5-T and cleft palate occurrence in Arkansas. Teratology 19:377-384

Neubert D, Dillmann I (1972) Embryotoxic effects in mice treated with 2,4,5-trichlorophenoxyacetic acid and 2,3,7,8-tetra-chlorodibenzo-p-dioxin. Arch Pharmacol 272:243-264

Organization for Economic Cooperation and Development (1981) Guidelines for testing of chemicals. Section 4:Health effects, 414, Teratogenicity

Ottolenghi AD, Haseman JK, Suggs F (1974) Teratogenic effects of aldrin, dieldrin, and endrin in hamsters and mice. Teratology 9:11-16

Palmer AK (1981) Regulatory requirements for reproductive toxicology: Theory and practice. In: Kimmel CA and Buelke-Sam J (eds) Developmental Toxicology. Raven Press, New York.

Pasi A, Embree, JW, Eisenlord GH, Hine CH (1974) Assessment of the mutagenic properties of diquat and paraquat in the murine dominant lethal test. Mutat Res 26:171-175

Rao KS, Burek JD, Murray FJ, John JA, Schwetz BA, Bell TJ, Potts WJ, Parker CM (1983) Toxicologic and reproductive effects of inhaled 1,2-dibromo-3-chloropropane in rats. Fundam Appl Toxicol 3:104-110

Rogers JM, Grabowski CT (1983) Mirex-induced fetal cataracts: Lens growth, histology and cation balance, and relationship to edema. Teratology 27:343-349

Schardein JL (1985) Chemically Induced Birth Defects - Volume 2 IN: DiCarlo FJ, Oehme FW (eds) Drug and Chemical Toxicology, Marcel Dekker Inc, New York

Schein LG, Thomas JA (1975) Effects of dieldrin on the uptake and metabolism of testosterone-1,2-^{3}H by rodent sex accessory organs. Environ Res 9:26

Schwetz BA, Sparschu GL, Gehring PJ (1971) The effect of 2,4-dichlorophenoxyacetic acid (2,4-d) and esters of 2,4-d on rat embryonal, foetal and neonatal growth and development. Food Cosmet Toxicol 9:801-817

Scotti TM, Chernoff N, Linder R, McElroy WK (1981) Histopathologic lens changes in mixrex-exposed rats. Toxicol Ltrs 9:289-294

Shepard TH (1986) Catalog of Teratogenic Agents Fifth Edition. The Johns Hopkins University Press, Baltimore and London

Sparschu GL, Dunn FL, Lisowe RW and Rowe VK (1971) Study of the effects of high levels of 2,4,5-trichlorophenoxyacetic acid on foetal development in the rat. Fd Cosmet Toxicol 9:527-530

Taylor JR, Selhorst, JB, Houff SA, Martinez AJ (1978) Chlordecone intoxication
   in man. 1. Clinical observations. Neurology 28:626-630

U.S. Environmental Protection Agency, Proceedings of Sponsored Conferences
   (1980) Assessment of risks to human reproduction and to development of
   the human conceptus from exposure to environmental substances. EPA-600/
   9-82-001

U.S. Environmental Protection Agency (1982) Pesticide Assessment Guidelines,
   Sub-division F, Hazard Evaluation: Human and domestic animals, EPA-540/
   9-82-025

Whorton D, Milby TH, Krauss RM, Stubbs HA (1979) Testicular function in
   DBCP exposed pesticide workers. J Occup Med 21:161-166

CARCINOGENICITY OF PESTICIDES

J.R.P. Cabral
International Agency
for Research on Cancer (IARC)
Lyon, France

INTRODUCTION

What is a pesticide? A pesticide is a substance capable of selectively killing a pest.

The pesticides (Table 1) are classed depending on the particular use intended: Herbicides are used for the control of a variety of weeds, Fungicides for the control of many plant fungal disease, Insecticides if insects are the target species and the Fumigants are used predominantly in control of nematodes in soil.

Table 1. Classification of pesticides

The importance of pesticides can be judged with data Table 2 and 3 published by Chapman (1978). Table 2 shows that between 1957 and 1976 the total market value of pesticides was up 330%. We can also see that in this same period the herbicides made a great leap forward controlling now almost half of the pesticide market, indicates clearly the decline of the organo-

NATO ASI Series, Vol. H13
Toxicology of Pesticides: Experimental, Clinical
and Regulatory Aspects. Edited by L. G. Costa et al.
© Springer-Verlag Berlin Heidelberg 1987

chlorine compounds, progressively substituted by the organophosphates, and the emergence of carbamates.

Table 2. The pesticide market at end user level for the world excluding Comecon countries ($ millions 1976 money).

	Market Value 1957	%	Market Value 1976	%
Insecticides	882	54	2395	34
Herbicides	255	16	3198	46
Fungicides	370	23	1086	15
Others	125	7	358	5
Total	1632	100	7037	100

Table 3. Breakdown of the insecticide market by class of product (%)

	1957	1976
Organochlorine	65	25
Organophosphate	25	53
Carbamate	--	16
Others	10	6

Now we come to another question and that is: why did I choose to rely on the _in vivo_ carcinogenicity effects of pesticides? First of all because the assays involving the induction of bacterial mutations show many limitations. Second - There are few epidemiological studies available. Reasons for this scarcity are: a) it is difficult to identify exposed groups of an adequate size which are not exposed to other chemicals. b) or in the case of highly polluting chemicals it may be difficult to identify non-exposed groups. This difficulty in the identification of non-exposed groups is however relatively marginal. Altogether carcinogenesis also follows the laws of dose-response and an epidemiological study on workers exposed to compounds like DDT can be good, even if the control population has clearly a minimal exposure to that pesticide.

In the absence of good human data, one has no choice but rely on experimental studies in laboratory animals as the only possible source of information.

The next question to be explored is: has the potential carcinogenicity of pesticides been investigated satisfactorily?

Here I would like to mention the data reported by the International Agency for Research on Cancer (IARC) between 1972 and 1986.

According to IARC the evidence that a chemical produces tumours in experimental animals is of two degrees:

a) Sufficient evidence for carcinogenicity is provided by experimental studies that show an increased incidence of malignant tumours in multiple species and following multiple routes and doses.

b) Limited evidence of carcinogenicity in animals because of inconclusive results.

It is important to stress that the concepts "sufficient evidence" and "limited evidence" indicate varying degrees of experimental evidence. They have to be always related to the knowledge available at a certain moment and they presuppose a continuous change with the acquisition of new data on the chemicals.

Otherwise one risks to have a rigid list containing the carcinogenic chemicals and all the rest, classified as "safe".

A total of 45 pesticides (a real small portion of all pesticides in commerce) have been so far evaluated by IARC with following observations: one is carcinogenic in man; in 15 there is sufficient evidence of carcinogenicity in laboratory animals; in 11 the evidence is limited; one is noncarcinogenic and in the remaining 17 pesticides the available studies are not adequate to make an evaluation of carcinogenic potential in animals or humans.

In Table 4 we have some examples of high volume pesticides (annual production greater than 2500 tons). Several of the pesticides will be considered in this review.

Table 4. Pesticides produced in quantities greater than 2500 tons per year.

ORGANOCHLORINE	ORGANOPHOSPHATE
BHC (technical)	Dichlorvos
Carbon tetrachloride	Carbamate
Chlordane	Carbaryl
2,4-D	Diallate
DBCP	Maneb
DDT	Thiram
Dieldrin	Zineb
Heptachlor	Ziram
Hexachlorobenzene	
Methoxychlor	
PCP	
2,4,5-T	
Toxaphen	

From IARC (1972-1986)

ORGANOCHLORINE PESTICIDES

DDT

DDT is present everywhere in the human environment and is stored by plants and animal tissues. As a consequence most food items are contaminated with it. In the U.S. an extensive survey on ready-to-eat food commodities carried our during 1964-1970 indicated that the daily intake of total DDT ranged between 0.0004 and 0.001 mg/kg/day, the higher values being antecedent to 1968 (Duggan and Corneliussen, 1972). In 1969, the maximum acceptable daily intake (ADI) established by FAO/WHO was 0.005 mg/kg/day.

Several studies in different countries carried out during 1951-1969 demonstrated that the concentration of DDT in human milk ranges between 0.05 and 0.2 ppm (Lang et al. 1951, Egan et al. 1965, Heyndrickx and Maes 1969). Assuming an average concentration of 0.1 ppm total DDT in milk, a newborn baby ingesting 700 g milk/day would have an intake in the order of 0.010.02 mg/kg/day. These intakes exceed, by 2-4 times, the DDT maximum ADI.

The concentration of DDT in cow's milk is definitely lower. The average daily intake in workers engaged in the manufacture of DDT has been estimated to be 0.2 mg/kg/day. There are already some indications available that following the restrictive legislation approved in many countries DDT contamination of food is showing a steady decrease (Duggan and Corneliussen, 1972).

The evidence of carcinogenicity of DDT can be discussed in the light of three types of data:

First. The concentration of DDT in tissues of terminal cancer patients. One study showed an almost identical average concentration of residues in human fat among 292 patients dying with cancer and 336 patients dying with other diseases (Hoffman et al., 1967). One other study (Radomski et al., 1968) showed that patients with non-neoplastic liver disease had fat and liver concentrations of DDT residues higher than in controls.

Second. Studies on groups professionally exposed to DDT. No cases of tumour or blood dyscrasia were found in two separate investigations concerning respectively 40 men followed for 0.4-8 years (Ortelee, 1958) and 63 men followed for at least 5 years (Laws et al., 1967). The daily intakes were calculated to average respectively 200 and 50 times those of the general population. The small number of workers studied, the short period of exposure and follow up render this study as of little relevance for assessing long-term risks of occupational exposure to DDT. There are no additional studies reported in the literature, designed to test the potential carcinogenicity of DDT in man.

Third. Long term studies in experimental animals.

The first reference to the carcinogenicity of DDT was made by Fitzhugh and Nelson in 1947. Technical DDT was fed to Osborne Mendel rats during 24 months at dietary levels ranging from 100-800 ppm. Among the 75 rats alive after 18 months, 4 had "low-grade" hepatic cellcarcinomas and 11 had hyperplastic nodules. This rat study has been recently re-evaluated by Reuber (1978). The review of the histology has shown a significant incidence of liver tumours, ranging from 36% to 54% in treated females and males. The incidence of liver tumours in controls was 0%.

Later studies by various authors with DDT in rats gave no evidence of carcinogenicity (Radomski et al. 1965) and others. Some of these studies were considered inadequate either because of low survival rate, insufficient data report or because the doses given were too low. More recently Rossi et al., 1977 observed a high incidence of liver-cell tumours in Wistar rats fed 500 ppm DDT for their lifespan.

In our own work with DDT in rats we have shown that concentrations of 125, 250 and 500 ppm DDT in the diet induced an increase in liver-cell tumours predominantly in the females (Cabral et al., 1978).

A study by Cabral and Shubik (1977) used Syrian Golden hamsters (Table 5). The animals were fed for life a diet containing 125, 250 and 500 ppm DDT. The results of this experiment confirm that the hamster is resistant to the carcinogenic effects of doses of DDT which are carcinogenic for other species. In the previous hamster experiment (Agthe et al., 1970) also did not given any signs of carcinogenicity.

In Table 6 we present a summary of long term-feeding studies with DDT: of a total of 18 experiments (9 were positive and 9 negative). The experimental evidence for carcinogenicity of DDT is actually based on the induction of liver-cell tumours in mice (Tomatis et al., 1972; Terracini et al., 1973) multigeneration studies. The studies in rats have provided contradictory results. This discrepancy, as well as the negative results with hamsters, suggest different metabolic pathways that we do not know yet the nature of. These could be important at the moment of extrapolation of the results to man.

Present knowledge on the carcinogenicity of DDT in animals does not

Table 5. Tumour incidence in hamsters given DDT.

Group	Effective no.	TBA No.	%	No. of tumours No.	per Hamster	More than one tumour No.	%	Adrenal No.	%	Liver tumours Liver-cell No.	%	Hemangio. No.	%	Other No.	%
Control	39 F	5	12.8	5	0.13	0	0	0	0	0	0	0	0	5	12.8
	40 M	3	7.5	3	0.08	0	0	3	7.5	0	0	0	0	0	0
DDT	28 F	3	10.7	3	0.11	0	0	0	0	0	0	0	0	3	10.7
	30 M	5	16.6	7	0.23	1	3.3	4	13.3	0	0	0	0	3	10.0
DDT 250	28 F	2	7.1	2	0.07	0	0	1	3.5	0	0	0	0	1	3.6
	31 M	8	25.8	11	0.35	3	9.6	6	19.3	1	3.2	2	6.4	2	6.5
DDT 500	40 F	8	20.0	13	0.33	3	7.5	3	7.5	0	0	0	0	10	25.0
	39 M	11	28.2	11	0.28	0	0	8	20.5	0	0	0	0	3	7.5

ANIMALS WITH

allow us to predict with certainty that DDT will not have a carcinogenic effect in man. There are still many gaps to fill before an overall evaluation can be made, but on the whole it seems that DDT presents a low risk.

Table 6. Summary of Long-term Feeding Experiments with DDT.

Species	No. experiments	Hepatocarcinogenicity	
		Positive	Negative
Mouse	8	6	2
Rat	7	3	4
Hamster	3	0	3

As a matter of fact a risk vs benefit evaluation of DDT was made by WHO (1973). Because of the DDT uses in world malaria control programmes, saving the lives of millions of people, it was concluded that the benefits obtained from the usage of DDT have surpassed its possible risk.

Heptachlor, Heptachlor epoxide and Chlordane

A comprehensive and up-to-date review on these pesticides has been recently published (IARC, 1979).

Heptachlor and heptachlor epoxide and chlordane were tested in mice and rats by administration in diet. Liver-cell carcinomas were reported in mice of both sexes following exposure to each of the three compounds.

No conclusions can yet be reached on the carcinogenicity of these chemicals in rats. A number of case reports have suggested a relationship between home use exposure to heptachlor or chlordane (either alone or in combination with other compounds) and blood dyscrasias (Infante et al. 1978). An association between both pre- and post-natal exposure to technical chlordane containing heptachlor and the development of neuroblastomas in children has been suggested, but this was based on a small absolute number of cases (Infante et al., 1978).

Dieldrin

Intake by the general population of dieldrin from food was reported to be between 0.00007 and 0.0001 mg/kg/day during the period 1964-1970 (Duggan and Corneliussen, 1972).

The epidemiological study in workers occupationally exposed to Dieldrin did not give evidence of a carcinogenic risk of Dieldrin to man, but this was based on a small number of workers followed up (Jager 1970; Versteeg and Jager 1973).

There have been at least 11 studies of carcinogenicity in experimental animals (Table 7). The ability of Dieldrin to produce liver-cell tumours in mice is evident. The data on rats have not provided conclusive evidence of carcinogenicity. Our own data in studies conducted in hamsters indicate that this species is resistant to the carcinogenic effects of doses of Dieldrin higher than those producing a very marked incidence of liver-cell tumours in mice.

Hexachlorobenzene (HCB)

Hexachlorbenzene has been used as a fungicide in various parts of the world, and is also an important impurity of several widely used pesticides, including Dacthal and PCNB. The ADI was withdraun by WHO/FAO in December 1978 (D. Clegg - Personal Communication). It is recognized that HCB residues also arise from sources other than its use as a fungicide. In fact the major source of HCB in the environment originates from the manufacture of many chlorinated hydrocarbons (Quinlivan et al. 1977).

One episode of epidemic poisoning in humans occurred in Turkey between 1955 and 1959 - Table 8.

The consumption of HCB-treated wheat seeds for at least one month caused an outbreak of toxic porphyria, which involved more than 3000 people, predominantly children. The mortality rate during these years ranged from 3-11% annually. In 1959 treatment of wheat with HCB was prohibited in Turkey. No new cases of the porphyric syndrome were thereafter reported. Howe-

Table 7. Summary of carcinogenicity experiments with dieldrin.

Species and Strains	Range of intakes (mg/kg bw/day)	Evidence of Carcinogenicity	Reference
Mice: C3HeB/fe	1.5	+	Davis & Fitzhugh (1962)
CF1	0.15 - 1.5	+	Walker et al. (1973)
CF1	1.5	+	Thorpe & Walker (1973)
B6C3F1	0.3 - 0.6	+ (in male mice)	NCI (1978b)
Rats: Carworth	0.12 - 1.2	-	Treon & Cleveland (1955)
Osborne-Mendel	0.02 - 7.5	+|	Fitzhugh et al. (1964)
Carworth	0.005 - 0.5	-	Walker et al. (1969)
Osborne-Mendel	1 - 2.5	-	Deichmann et al. (1970)
Osborne-Mendel	1.5 - 3.2	-	NCI (1978b)
Fischer 344	0.1 - 2.5	-	NCI (1978c)
Hamsters:Syrian golden	1.7 - 15	-	Cabral et al. (1979)

ver many undesirable symptoms (skin hyperpigmentation, hirsutism, thyroid and liver enlargement etc.) due to this toxic porphyria persist still today (Peters et al., 1978).

Table 8. Epidemic of poisoning by hexachlorobenzene in Turkey (1955-1959).

Kind of Accident	Eating formulation
Material Contaminated	Seed grain
No. of Cases	Over 3000 (mostly 4-15 year old children)
Mortality	3-11% Annually (1955-1959)
	95% Among breast-fed infants
Symptoms of HCB-induced porphyria	Cutaneous, hepatic and neurological
References	Cam and Nigogosyan (1963), Dogramaci (1964), Elder (1978), Peters et al. (1966) and Peters (1976).

Experimental evidence on the carcinogenicity of HCB was first reported by Cabral et al. 1977 - Table 9 shows the tumour incidence in hamsters fed for life a diet containing 50, 100 and 200 ppm HCB.

The percentage of tumour bearing animals (TBA) varied, with controls showing a 10% incidence and hamsters treated with 200 ppm HCB a 92% incidence. There was, therefore, an obvious difference relative to treatment.

All 15 thyroid tumours were present only in treated animals. There was a significant increase ($P < 0.05$) in the occurrence of these tumours in males treated with 200 ppm HCB. Also for liver-cell tumours you can notice a dose-response effect. Liver haemangioendotheliomas - these vascular tumours were observed only in HCB - treated groups. There is a doseresponse

Table 9. Tumour incidence in hamsters given HCB.

Group	Effective No.	Sex	TBA No.	TBA %	No. of tumours No.	per Hamster	More than one tumour No.	More than one tumour %	ANIMAL WITH Thyroid No.	Thyroid %	Liver-cell Tumour No.	Liver-cell Tumour %	Hemangio. Liver No.	Hemangio. Liver %	Spleen No.	Spleen %	Other No.	Other %
Control	39	F	5	12.8	5	0.13	0	0	0	0	0	0	0	0	1	2.5	4	10.2
	40	M	3	7.5	3	0.08	0	0	0	0	0	0	0	0	0	0	3	7.5
HCB50	30	F	16	53.3	21	0.70	4	13.3	2	6.6	14	46.6	0	0	0	0	5	16.6
	30	M	18	60.0	27	0.90	8	26.6	0	0	14	46.6	1	3.3	1	3.3	11	36.6
HCB100	30	F	18	60.0	32	1.06	11	36.6	1	3.3	17	56.6	2	6.6	3	10.0	9	30.0
	30	M	27	90.0	45	1.50	14	46.6	1	3.3	26	86.6	6	20.0	3	10.0	9	30.0
HCB200	60	F	52	86.6	73	1.21	15	25.0	3	5.0	51	85.0	7	11.6	4	6.6	8	13.3
	57	M	56	98.2	87	1.52	27	47.3	8	14.0	49	85.9	20	35.0	4	7.0	6	10.5

Table 10. Incidence, size, multiplicity and latency of liver-cell tumours (LCT) in hamsters exposed to HCB.

| Group | Hamsters with | Size of nodes (mm) | | | Single | Multiplicity Multiple | No. nodes per | Latency (whs) | |
%	%	4-5 %	6-10 %	10 %	%	%	hamster	Range	Average
Control F	0	0	0	0	0	0	0	0	0
M	0	0	0	0	0	0	0	0	0
HCB50 F	46.6	42.9	35.7	21.4	35.7	64.3	2.4	(46-107)	71.9
M	46.6	0	57.1	42.9	14.3	85.7	4.4	(58-114)	82.7
HCB100 F	56.6	17.6	41.2	41.2	17.6	82.4	3.3	(46- 92)	73.3
M	86.6	11.6	34.6	53.8	15.4	84.6	4.0	(44-101)	78.3
HCB200 F	85.0	8.0	46.0	46.0	10.0	90.0	4.7	(18- 80)	38.5
M	85.9	6.1	38.8	55.1	8.2	91.8	4.2	(31- 89)	66.5

effect. Three of these tumours appearing in males treated with HCB 200 metastasized. Vascular tumours were also observed in the spleen.

Table 10 gives information on some biological aspects of the liver-cell tumours induced by HCB.

In the HCB groups the liver-cell tumour (LCT) incidence increased from 46% in HCB 50 to 85% in HCB 200. We can further see that there is a dose-response effect for size on the LCT, multiplicity and latency at death - note that the first LCT appeared after 18 weeks of treatment. So, this seems to be a rapidly progressive situation, although no metastases have been seen.

Table 11 summarizes, the long-term feeding experiments with HCB. I have just illustrated the results of our hamster study. In mice, our results show that concentrations of 100 ppm or more of HCB in the diet induced liver-cell tumours in mice of both sexes (Cabral et al., 1979).

Our findings that HCB is carcinogenic in two species, and preliminary daa showing same effect in rats, provide sufficient evidence of carcino-genicity to recommend greatest caution on the use of this material.

Table 11. Summary of long term feeding experiments with HCB.

Species	No. of Animals		Intake	Effect on	Evidence of
	Control	Treated	mg/kg/day	Survival	Carcinogenicity
Mice	100	200	6-24	Yes	Carcinogenicity for the liver
Hamsters	80	239	4-16	Yes	Significant induction of liver-cell tumours liver haemangioendo-thelioma and thyroid adenoma. Dose-response relationship.

In addition investigations on long-term effects in humans known to have been exposed to Hexachlorobenzene seem feasible and it is remarkable that they have not been reported or even carried out.

If we now compare the carcinogenicity of DDT with that of HCB (Table 12) we can see that the latter poses many more problems. In the case of HCB there is a wide spectrum of carcinogenic activity for different tissues and different animal species. Further, we have also observed an increase in the number of TBA, a marked shortening of the lifespan and a reduced latency period for onset of liver-cell tumours. With DDT we have clear cut hepatocarcinogenic results only in mice and this without shortening of lifespan. In hamsters we have noted the marked resistance of that species.

Table 12. Comparative carcinogenicity of HCB and DDT.

	Rat	Hamster	Mouse
HCB	+	−	+
DDT	±	−	+

ORGANOPHOSPHATES

Studies on the carcinogenicity of this family of pesticides are very scarce. This scarcity of data is astonishing especially if one considers the growing importance of these compound as alternatives to the organochlorine pesticides.

Some of the most used OP compounds are malathion, parathion, dichlorvos, diazinon, disulfoton, etc. As an example we will consider here a well-known member of this group: dichlorvos. This compound has a very wide range of uses in agriculture, in public health programmes of pest control and in household products. An acceptable daily intake (ADI) of 0.004 mg/kg bw was granted by WHO/FAO in December 1974 (WHO, 1975). Studies have detected residues of dichlorvos in cooked foods at levels of 0.01-0.03 mg/kg - Elgar et al., 1972. There are no epidemiological studies available. However there are reports in the literature of groups of people, babies and newborns, intentionally exposed to dichlorvos (Cavagna et al., 1969; Vigliani, 1971) that would be suitable for epidemiological follow up.

Dichlorvos has been shown to be an alkylating agent _in vitro_ and also to have mutagenic effects in bacterial systems. Despite this evidence, dichlorvos has not found to be carcinogenic in experimental studies, as is shown in the Table 13.

Table 13. Results long-term feeding studies with Dichlorvos.

Species	Range intake (ppm)	Effect survival	Length exposure (months)	Carcinogenicity	Reference
Mice	318-635	Yes	20	None	NCI 1977b
Rats	150-326	No	20	None	NCI 1977b

Dichlorvos was give to mice and rats at MTD levels for a period of 20 months (NCI 1977 b). No statistically significant excess of tumours was observed in treated mice and rats compared with respective controls. How-ver, in treated mice few esophageal tumours were reported (NCI 1977 b). This is another example of experimental results that one is not in the condition of evaluating their significance.

It is true that dichlorvos is rapidly metabolised _in vivo_. However, also nitrosamines which are powerful alkylating agents and rapidly metabo-lised, can cause and they produce cancer when administered (single shot) to newborn and young animals (Terracini and Magee 1964).

CARBAMATES

Carbamates, relative newcomers to pesticidal use, act as the organophosphates principally by inhibition of acetylcholinesterases. One important difference between OP compound and the carbamates is that the inhibition of cholinesterase by the latter is reversible. The carbamates also do not cause the delayed neurotoxicity that has been demonstrated with some organophosphates (Johnson and Lauwerys 1969).

As in the case of OP compounds there is little information available on the potential carcinogenicity of carbamate pesticides.

Innes et al. (1969) (Table 14) surveyed the carcinogenicity of these compounds. In mice, Diallate, increased the incidence of liver-cell tumours. The other important carbamate pesticides considered here did not show any significant effects. Andrianova and Alekseev (1970) used rats to study the long-term effects of Carbaryl, Maneb, Zineb and Ziram in rats. Only Ziram appeared to give some very marginal indication of carcinogenicity in the rat.

Table 14. Results of long-term feeding studies in mice with carbamate pesticides.

Compound	Intake (ppm)	Effect Survival	Carcinogenicity
Carbaryl	14	No	None
Diallate	560	No	Increased incidence of liver-cell tumours
Maneb	158	No	None
Thiram	26	No	None
Zineb	1298	No	None
Ziram	15	No	None

Innes et al. 1969

However some interesting challenges from this family of compounds have become evident.

First - the problem of impurities and of the metabolic and degradation products of thiocarbamates. Some of these compounds (Maneb, Zineb, Thiram, etc.) create some concern because of their goitrogenic (thyroid hyperplasia inducers) effects in laboratory animals (Blackwell-Smith et al., 1953).

It is possible that these effects are due to ethylene thiourea (ETU). ETU has been shown to be a process impurity (Bontoyan et al., 1972), a breakdown product in storage (Petrosini et al. 1963) and probably also a breakdown product in vivo. ETU is known to produce tumours in the thyroid and other sites in mice (Innes et al., 1969) and rats (Ulland et al., 1972).

Second - there are recent reports that many alkylcarbamates and thiocarbamates by reaction with nitrite are capable of producing N-nitroso compounds in vitro and in vivo systems.

N-nitrosocarbaryl a derivative of carbaryl, a widely used pesticide, is a potent carcinogen in rats (Eisenbrand et al., 1975, 1976; Lijinsky and Taylor 1976). Since nitrite is present in many foods and carbamate residues have also been reported in a wide range of foods, it seems that the conditions for the formation of N-nitroso derivatives in man could take place in real life. The extrapolation of the results of carbamate nitrosation to man is a very complicated one.

However, this kind of data suggests the need of appropriate investigations to study any possibilities of large scale formation of N-nitroso compounds.

REFERENCES

Agthe, C. et al. (1970) Study of the potential carcinogenicity of DDT in Syrian Golden hamsters. Proc. Soc. Exp. Med. (N.Y.) 134, 113.

Andrianova, M.M. and Alekseev, I.V. (1970) On the carcinogenic properties of the pesticides Sevin, Maneb, Ziram and Zineb. Vop. Pitan., 29, 71.

Blackwell-Smith, R. et al. (1953) Toxicologic studies on zinc and disodium ethylene bisdithiocarbamates. J. Pharmacol. Exp. Ther., 109, 159.

Bontoyan, W.R. et al (1972) Survey of ethylenethiourea in commercial ethylene bis-dithiocarbamate formulations. J. Ass. Off. Analyt. Chem. 55, 923.

Cabral, J.R.P. and Shubik, P. (1977) Lack of carcinogenicity of DDT in hamsters. Federation Proc. 36, 1086.

Cabral J.R.P. et al. (1977) Carcinogenic activity of hexachlorobenzene in hamsters. Nature, 269, 510.

Cabral, J.R.P. et al. (1978) Effects of long-term DDT intake in rats. European Society of Toxicology, 20th Annual Meeting, West Berlin.

Cabral, J.R.P. et al. (1979) Carcinogenesis of hexachlorobenzene in mice. Int. J. Cancer., 23, 47.

Cabral, J.R.P. et al. (1979) A carcinogenicity study of the pesticide dieldrin in hamsters. Cancer Letters, 6, 241.

Cam, C. and Nigogosyan, G. (1963) Acquired toxic porphyria cutanea tarda due to hexachlorobenzene. J. Amer. Med. Ass. 183, 88.

Cavagna, G. et al. (1969) Clinical effects of exposure to DDVP (Vapona) insecticide in hospital wards. Arch. Environ.Hlth. 19, 112.

Chapman, T. (1978) Pesticides - A period of progress. Span 21, 62.

Davis, K.L. and Fitzhugh, O.G. (1962) Tumorigenic potential of aldrin and dieldrin for mice. Toxicol. Appl. Phamacol., 4, 187.

Deichmann, W.G. et al. (1970) Tumorigenicity of aldrin, dieldrin and endrin in albino rat. Industr. Med. Surg., 39, 426.

Dogramaci, I. (1964) Porphyrias and porphyrin metabolism with special reference to porphyria in childhood. Advanc. Pediat. 13, 11.

Duggan, R.E. and Corneliussen, P.E. (1972) Dietary intake of pesticide chemicals in the United States (III), June 1968 - April 1970. Pest. Monit. J., 5, 331.

Egan, H. et al. (1965) Organchlorine pesticide residues in human fat and human milk. Brit. Med. J., iii, 66.

Eisenbrand, G. et al. (1975) The reaction of nitrite with pesticides. II. Formation, chemical properties and carcinogenic activity of the N-nitroso derivative of N-methyl-1-naphthyl carbamate (Carbaryl). Fd. Cosmet. Toxicol., 13, 365.

Eisenbrand, G. et al. (1976) Carcinogenicity in rats of high oral doses of N-nitrosocarbaryl, a nitrosated pesticide. Cancer Letters, 1, 281.

Elder, G.H. (1978) Porphyria caused by hexachlorobenzene and other polyhalogenated aromatic hydrocarbons. In "Heme and Hemoproteins Handbook of Experimental Pharmacology", vol. 44 (Edited by De Matteis, F. and Aldridge, W.N.) Springer-Verlag, Berlin, p. 157.

Elgar, K.E. et al. (1972) Dichlorvos (2,2-dichlorovinyl dimethyl phosphate) residue in food arising from the domestic use of dichlorvos PVC (Polyvinyl chloride) strips. Pestic. Sci., 3, 601.

Fitzhugh O.G. and Nelson, A.A. (1947) The chronic oral toxicity of DDT 2,2-bis p-chlorophenyl -1,1,1-trichloroethane). J. Pharmacol. Exp. Ther. 89, 18.

Fitzhugh, O.G. et al. (1964) Chronic oral toxicity of Aldrin and Dieldrin in rats and dogs. Fd. Cosmet. Toxicol., 2, 551.

Heyndrickx, A. and Maes, R. (1969) The excretion of chlorinated hydrocarbon insecticides in human mother milk. J. Pharm. Belg., 24, 459.

Hoffman, W.S. et al. (1967) Relation of pesticide concentrations in fat to pathological changes in tissues. Arch. Environ. Hlth., 15, 758.

I.A.R.C. (1972-1986) Monographs of the evaluation of carcinogenic risk of chemicals to man, vol. 1-39. Lyon, France: Internatioal Agency for Research in Gancer.

Infante, P.F. et al. (1978) Blood dyscrasias and childhood tumours and exposure to chlordane and heptachlor. Scand. J. Work Environ. and

Health, 4, 137.

Innes, J.R.M. et al. (1969) Bioassay of pesticides and industrial chemicals for tumorigenicity in mice. A Preliminary Note. J. Nat. Cancer. Inst., 42, 1101.

Jager, K.W. (1970) Aldrin, Dieldrin, Endrin and Telodrin. An epidemiological and toxicological study of long-term occupational exposure. (London-Elsevier).

Johnson, M.K. and Lauwerys, R. (1969) Protection by some carbamates against the delayed neurotoxic effects of di-isopropyl phosphorofluoridate. Nature, 222, 1066.

Lang, E.P. et al. (1951) Occurrence of DDT in human fat and milk. Arch. Industr. Hyg., 3, 245.

Laws, E.R. et al. (1967) Men with intensive occupational exposure to DDT. A clinical and chemical study. Arch. Environ. Hlth., 15, 766.

Lijinsky, W. and Taylor, H.W. (1976) Carcinogenesis in Sprague-Dawley rats by N-nitroso-n-alkylcarbamate esters. Cancer Letters, 1, 275.

N.C.I. (National Cancer Institute) (1977b) Bioassy of dichlorvos for possible carcinogenicity, Technical Report Series No. 10, DHEW Publication No (NIH) 77-810.

N.C.I. (National Cancer Institute) (1978b) Bioassyas of Aldrin and Dieldrin for possible carcinogenicity, Technical Report Series No. 21, DHEM Publication No (NIH) 78-821.

N.C.I. (National Cancer Institute) (1978c) Bioassay of Dieldrin for possible carcinogenicity, Technical Report Series No. 22, DHEM Publication No. (NIH) 78-822.

Ortelee, M.F. (1958) Study of men with prolonged intensive occupational exposure to DDT. Arch. Industr. Hlth., 18, 433.

Peters, H.A. et al. (1966) Hexachlorobenzene-induced porphyria: effect of chelation on the disease, porphyrin and metal metabolism. Amer. J. Med. Sci., 251, 104.

Peters, H.A. (1976) Hexachlorobenzene poisoning in Turkey. Fed. Proc. 35, 2400.

Peters, H.A. et al. (1978) Porphyria 20 years after hexachlorobenzene exposure. Neurology, 28, 333.

Petrosini, G. et al. (1963) Modifications of dithiocarbamate fungicides during storage. Notiz. Mal. Piante, 65, 9.

Quinlivan, S.C. et al. (1977) Sources, characteristics and treatment and disposal of industrial wastes containing hexachlorobenzene. J. Haz. Mat., 1, 343.

Radomski J.L. et al. (1965) Synergism among oral carcinogens. I. Results of simultaneous feeding of four tumorigens to rats. Toxicol. Appl. Pharmacol., 7, 652.

Radomski J.L. et al. (1968) Pesticide concentrations in the liver, brain and adipose tissue of terminal hospital patients. Fd. Cosmet. Toxicol., 6, 209.

Reuber, M.D. (1978) Carcinomas of the liver in Osborne-Mendel rats ingesting DDT. Tumori, 64, 571.

Rossi, L. et al. (1977) Long-term administration of DDT or pheno-barbitalNa in Wistar rats. Int. J. Cancer, 19, 179.

Terracini, B. and Magee, P.N. (1964) Renal tumours in rats following injection of dimethylnitrosamine at birth. Nature, 202, 502.

Terracini, B. et al. (1973) The effects of long-term feeding of DDT to BALB/c mice. Int. J. Cancer, 11, 747.

Thorpe, E. and Walker, A.I.T. (1973) The toxicology of Dieldrin (HEOD). II. Comparative long-term oral toxicity studies in mice with Dieldrin, DDT, Phenobarbitone, ß-BHC and -BHC. Fd. Cosmet. Toxicol., 11, 433.

Tomatis, L. et al. (1972) The effect of long-term exposure to DDT on CF1 mice. Int. J. Cancer, 10, 489.

Treon, J.F. and Cleveland, F.P. (1955) Toxicity of certain chlorinated hydrocarbon insecticides for laboratory animals with special reference to Aldrin and Dieldrin. J. Agric. Fd. Chem., 3, 402.

Ulland, B.M. et al. (1972) Thyroid cancer in rats from ethylene thiourea intake. J. Nat. Cancer Inst., 49, 583.

Versteeg, J.P.J. and Jager, K.W. (1973) Long-term occupational exposure to the insecticides Aldrin, Dieldrin, Endrin and Telodrin. Brit. J. Industr. Med., 30, 201.

Vigliani, E.C. (1971) Exposure of newborn babies to Vapona insecticide. Toxicol. Appl. Pharmacol., 15, 345.

Walker, A.I.T. et al. (1969) The toxicology and pharmacodynamics of Dieldrin (HEOD): two-year oral exposure of rats and dogs. Toxicol. Appl. Pharmacol., 15, 345.

Walker, A.I.T. et al. (1973) The toxicology of Dieldrin (HEOD). I. Longterm oral toxicity studies in mice. Fd. Cosmet. Toxicol., 11, 415.

WHO (1973) Safe use of pesticides. WHO Techn. Rep. Ser., No. 513.

WHO (1975) 1974 Evaluations of some pesticide residues in food. WHO Pest. Res. Ser., No. 4.

DERMAL TOXICITY OF PESTICIDES

Corrado L. Galli, Marina Marinovich
Cosmetological Biology and Toxicology Centre
Institute of Pharmacological Sciences,
University of Milan,
Via Balzaretti 21, Milan
Italy

Incidence of skin illnesses on total illnesses.

In 1980 the Bureau of Labor Statistics compiled 5 million reports of work-related illnesses and injury regarding 75 million people employed. Skin disease, the wast majority of which is attributable to effects of toxic chemicals, accounts for greater than 40% of all reported occupational disease (Table 1). Manufacturing, a sector that employs 26% of working population, accounted for 60% of reported skin diseases. Agriculture, forestry and fishing (1.3% of the total number of workers) accounted for 5% of reported skin diseases.

These data drawn from worker report from all 50 USA states exclude self-employed individuals, farmers with less than 11 employers and employers of federal, state and local government agencies (U.S. Dept. of Labor. 1980). The true incidence is suspected to be ten to fifty times greater than the reported incidence. These disorders cause a serious personal and economic burden in lost production and wages, increased medical costs and ill health.

It has been calculated that in 1970 perhaps from 20 to 30 million dollars of work costs were lost annually due to occupational skin disease (O.S.H.A. 1978).

Skin as site of absorption of pesticides.

General considerations.

Recent studies have enphasized the skin as a prime route of absorption for

NATO ASI Series, Vol. H13
Toxicology of Pesticides: Experimental, Clinical
and Regulatory Aspects. Edited by L. G. Costa et al.
© Springer-Verlag Berlin Heidelberg 1987

Table 1. Occupational skin diseases in United States (1980).

	Total illnesses	Skin illnesses	Skin illnesses × employment / Total skin illnesses ×100
Manufacturing	76100	33800	60.1
Services	19000	8200	14.6
Wholesale and retail trade	12200	4300	7.6
Transportation and public utilities	8000	3300	5.9
Construction	7800	3000	5.3
Agriculture, forestry and fishing	4200	2800	5.0
Mining	1500	400	0.7
Finance, insurance and retail trade	1500	400	0.7
Total	130200	56200	

Modified from S.M. Worobec and J.P. DiBeneditto, in Cutaneous Toxicity eds. V.A. Drill, P. Lazar, Raven Press, N.Y., 1984.

certain occupational and environmental chemicals. The main barrier to the penetration of pesticides and other substances is the stratum corneum of the epidermis: it has been proposed that this layer functions as a two compartments system. The lipophilic compartment, is composed mostly of lipids and is the pathway through which nonpolar molecules diffuse. The intracellular material, namely proteins, is the compartment through which water and polar molecules diffuse.

The flux of water through the horny layer is a function of its water content. The higher its water content the more easily molecules can move through the horny layer.

Increasing attention is being paid to the lipid compartment as the part of the horny layer that nonpolar molecules such as the steroids dissolve, thereby allowing this tissue to serve as a reservoir.

For strongly lipid-soluble molecules that enter the horny layer easily it is evident that the rate-limiting barrier is not the horny layer but the lipid/aqueous fluid interface at the base of the horny layer.

Vehicles influences skin penetration by altering the partition coefficient. In addition a vehicle can enter the stratum corneum and alter its barrier characteristics.

Table 2. Factors affecting percutaneous absorption.

Concentration of applied dose ($\mu g/cm^2$)
Surface area of applied dose (cm^2)
Total dose
Application frequency
Duration of contact
Site of application
Temperature
Vehicle
Substantivity (nonpenetrating surface adsorption)
Wash-and-rub resistance
Volatility
Binding
Individual and species variations
Skin condition
Occlusion
Sources: Wester and Maibach (1983).

Direct and indirect factors (table 2) may affect the absorption: indirect factors include race, age, sex, skin condition and injuries, personal hygiene and allergy, temperature and humidity of the workplace. For example infant skin is more permeable than is adult skin which explains some fatalities among children treated for lice with malathion or lindane.

Temperature and humidity of the work areas, unconfortable protective clothing and the resulting skin perspiration can increase the adverse effect of certain pesticides.

Absorption is also dependent on the site of exposure (table 3). Absorption is more complete after the pesticide is applied to the face, scalp, neck, axilla and scrotum (Maibach et al. 1971). Skin areas where the follicles are numerous or with disrupted or chemically damaged epidermal barrier are much more permeable.

Direct factors involved in the dermal toxicity of pesticides include the chemical structure of the compounds, the extent of surface areas exposed, duration and persistence of exposure.

Either polar and nonpolar pesticides can diffuse through the horny layer by different molecular mechanism, but generally lipid-soluble pesticides are more readily absorbed.

A too low or high pH of the skin enhances the transepidermal entry of pesticides. Vehicles influence skin penetration by altering the partition coefficient and by affect the barrier characteristics of the horny layer.

Finally, a definite increase of the absorption occurs with enhancement of the contaminated skin surface as well as with the prolonged time of contact. Hexane hand rinsings from workers have been analysed to determine the persistence on the hands of workers. The persistence is 2 years for chlordane and dieldrin, 7 days for methoxychlor, captan and malathion, 1-112 days for endosulfan, DDT, Kelthane, parathion.

The skin as an organ of accumulation of environmental chemicals is a relatively new concept. Experimental studies demonstrate that the skin is a major tissue storage site for PCB and polybrominated biphenyls.

Moreover the skin actively metabolized chemicals, drugs and toxic compounds as PCBs, TCDD, chlordane, glucocorticoids, PAH. Modifications in skin metabolism can occur as a result of enzyme induction and to be influenced in certain disease (e.g. psoriasis) and by heredity (Bickers et al. 1978, Pannatier et al. 1978, Parkinson et al. 1980, Fouts 1982).

Table 3. In vivo percutaneous absorption of pesticides in different anatomic regions in human.

Anatomic region	Percent dose absorbed[a]	
	Parathion	Malathion
Forearm	8.6	6.8
Palm	11.8	5.8
Foot	13.5	6.8
Abdomen	18.5	9.4
Hand dorsum	21.0	12.5
Forehead	36.3	23.2
Axilla	64.0	28.7
Jaw angle	33.9	69.9
Fossa cubitalis	28.4	---
Scalp	32.1	---
Ear canal	46.6	---
Scrotum	101.6	---

From Maibach and Feldman (1974).

[a] Per 24-h exposure at concentration of 4 $\mu g/cm^2$.

Systemic toxicity.

Pesticides are biologically active chemicals and are unfortunately no specific for the target. It is therefore essential that the potential risk involved with the usage of a pesticide are carefully assessed and on decision made regarding the suitability of the pesticide for registration.

Exposure is a broad term and may have many different meanings. In the manufacturing workplace the concentration of the chemical in ambient air is usually measured, and this exposure estimate is used to evaluate hazard. In an outdoor environment, such as that of pesticide applicators, the dermal route of exposure has been established as the predominant one for organophosporus pesticides (Wolfe et al. 1967, 1972; Durham 1972) accounting for 87% of the total exposure. Percutaneous absorption is reported as the probable route of entry in 65-85% of all cases of occupational poisoning with pesticides. The penetration of toxic chemicals through the cutaneous barrier and their subsequent distribution throughout the body can cause local toxic effect and systemic poisoning.

The effects depend on the toxic action of the substance and on the amount and rate at which it is absorbed from the application situ.

The spraying or dusting of pesticides has been shown to result in deposition on exposed skin of 20-1700 times the amount reaching the respiratory tract (Feldman et al. 1974).

The traditional method for estimating dermal contact exposure is the attachment of absorbent patches to exposed skin surfaces and the outside of the clothing (Durham et al. 1962). The concentration of pesticide on the patch was then extrapolated to the appropriate surface area of the body.

The limitations of patch data are the variability of the results (Franklin et al. 1981) and that the measured levels give no indication of the amount of pesticide that is absorbed and ultimately available to exert a toxic effecton the target tissue (Franklin 1984). More recently has been demonstrated that the analysis of urinary metabolites represents a suitable methodology to obtain an indication of the exposure to pesticides.

For example strong correlation (fig. 1) was observed in orchard workers between the amount of azinphos-methyl sprayed and the 48-h dimethylthiophosphate metabolite in their urine.

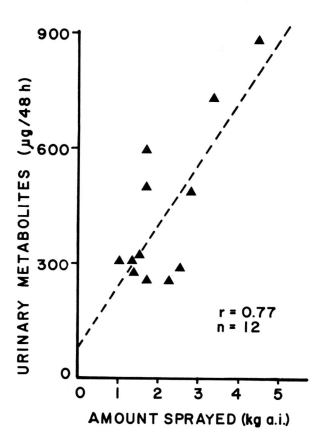

Fig. 1 Urinary metabolites of azinphos-methyl in orchard workers as a function of the amount of azinphos-methyl sprayed expressed in terms of kilograms of active ingredient (Kg a.i.). From Franklin (1984).

Although the presence of metabolites in the urine indicates the worker was exposed to the pesticide, it is difficult to correlate these levels quantitatively with the amount of pesticide impinged on the skin. Numerous studies done in rats and man have shown a relationship between exposure to

Table 4. Comparison of Acute Oral and Acute Dermal Toxicities of Various Insecticidal Compounds to Female Rats.

	Oral mg/Kg	Dermal mg/Kg	Dermal/oral ratio
Chlorinated hydrocarbons			
Aldrin	60	98	1.63
Chlordane	430	530	1.23
DDT	118	2510	21.27
Dieldrin	46	60	1.30
Endrin	7.5	15	2.0
Heptachlor	162	250	1.54
Isodrin	7.0	23	3.29
Kelthane (dicofol)	1000	1000	1.00
Lindane	91	900	9.89
Toxaphene	80	780	9.75
Organophosphates			
Chlorthion	980	4100	4.18
DDVP (dichlorvos)	56	75	1.34
Delnav	23	63	2.74
Demeton	2.5	8.2	3.28
Diazinon	76	455	3.79
Dicapthon	330	1250	5.99
Dipterex	560	2000	3.57
EPN	7.7	25	3.25
Azinphosmethyl (Guthion)	11	220	20.00
Malathion	1000	4444	4.44
Methilparathion	24	67	2.79
Parathion	3.6	6.8	1.89
Schradon	42	44	1.05
Thimet	1.1	2.5	2.27
Trithion	10.0	27	2.70
Others			
Isolan	13	6.2	0.48
Carbaryl	500	4000	8.0
Calcium arsenate	298	2400	8.05
Lead arsenate	1050	2400	2.29
Nicotine sulfate	83	285	3.43

From Gaines (1960). Stomach tube with peanut oil vs. dermal tests with xylene applied on 3.0 by 4.5 cm rectangles of shaven backs.

organophosphorus pesticides and alkyl phosphate excretion (Skafik et al. 1973, Bradway et al. 1977, Franklin et al. 1983).

Accurate estimates of exposure are an essential component of risk assessment. Patch data are poor predictors of exposure in man. Therefore the relationship between dermal dosage and urinary metabolite data would provide a mechanism to express metabolite data in terms of dermal dosage. The incidence of occupational poisoning from nearly all toxic insecticides has been found to be more closely related to the acute dermal LD_{50} than to the acute oral LD_{50} in animals. The relation between acute oral and acute dermal toxicity is shown in table 4.

The WHO has recommended a classification of pesticides according to hazard (table 5) (WHO 1975).

Table 5 Acute dermal LD_{50} for rats (mg/kg body weight).

Class of hazard	According to WHO (18)	
	Solids[a]	Liquids
Extreme	Ia[b] 11	Ia 41
High	Ib 10-100	Ib 40-400
Moderate	II 100 -1000	II 400-4000
Slight	III 1000	III 4000

From Bainova (1980).

[a] The terms "solid" and "liquid" refer to the physical state of the active ingredient or the formulation being classified.

[b] The number of the class according to the respective classification.

An other index of percutaneous pesticide toxicity (Kundiev 1975, Hudson et al. 1979) it has been proposed:

$$\text{index of percutaneous risk} = \frac{LD_{50} \text{ oral}}{LD_{50} \text{ dermal}} \times 100$$

Type of chemically induced skin injury.

Contact dermatitis.

Contact dermatitis is still by far the most frequent occupational skin disease. Data indicate that in 1977, 92.5% of the cases of occupational skin disease were contact dermatitis (Baginsky 1978), either irritant or allergic.

Allergic contact dermatitis is generally an expression of cell mediated delayed hypersensitivity. It is caused by low molecular weights chemicals capable of transepidermal penetration and of forming covalent bonds with carrier proteins. These substances are called haptens and become complete allergens only after conjugation with proteins of skin and blood.

Ultraviolet hight (UVA and UVB) has been found to influence the development of contact dermatitis, as well as the timing and route of exposure to allergenes (Morison et al. 1981).

New information has developed on modulation of mitogenic response to human T cells in prostaglandin and leukotriene research, and its possible role in allergic contact dermatitis (Stobo et al. 1979).

Contact urticaria

Certain substances on contact with intact skin can produce localized urticaria but also generalized cutaneous eruptions and extra cutaneous symptoms. This contact urticaria may have nonimmunological or immunological mechanism.

Protein contact dermatitis or atopic contact dermatitis is a new category of allergic contact dermatitis.

Pigmentary disorders

Occupationally related pigmentary disorders can be classified into three categories: discoloration or staining increase or decrease of melanization. Increase in melanization is by far the most frequent pigmentary problem. First and second degree burns as well as dermatitis often resolve with permanently hyperpigmented areas. Hypopigmentation can occur after third-degree burns or bullous dermatoses.

Chemicals as phenol, catechol, and their derivatives are inducers of pigment loss. Anatomic alterations in melanosomes and melanin biosynthesis have been clarified.

Mice and guinea pig models have been recommended for predictive screening for depigmentation produced by topically applied chemicals (Gellin et al. 1979, Yonemoto et al. 1983).

Occupational acne

Several types of acneiform eruptions can result from occupational exposure. Its microscopic features resemble those of acne volgaris: hyperkeratinization of the sebaceous gland duct and follicle opening with plugging of the pore opening.

The distribution of these lesions, however, follows the areas of greatest contact with the compounds, namely the backs of the hands and forearms.

A more serious form of occupational acne is chloracne or halogen acne since it has been associated with both chlorinated and brominated aromatic compounds. Halogen acne is an acneiform disorder of the pilosebaceous glands which is persistent, refractory to standard acne treatments, and associated with systemic toxicity. Prior to World War II, reported episodes involved exposure to chloronaphthalenes and diphenyls. Since then, other chloracnegens have been identified. Currently, public concern is focused on exposure to 2,3,7,8-tetrachlorodibenzo-p-dioxin (TCDD) which is an intermediate in the synthesis of 2,4,5-trichlorophenol (an herbicide) and hexachlorophene. About 20 episodes of TCDD exposure have occurred through industrial accidents. The exposures have usually been to trichlorophenol, TCDD, and sodium hydroxide in some combination; the initial skin reactions have been those of a chemical burn. Chloracne develops weeks to months later, sometimes accompanied by porphyria cutanea tarda, hyperpigmentation and hypertrichosis without porphyria, central and peripheral nervous system disorders, liver, kydney, and pancreatic disorders, psychiatric disturbances, hyperlipidemia and/or hypercholesterolemia. Increased incidence of depressed and abnormal sperm counts was detected in one group of 47 railroad workers exposed to TCDD, phenol, and chlorophenols.

The studies indicate that chloracnegens vary in their biological effects and that genetic factors play a role in determining their final toxic effects.

Bromination renders a compound more acnegenic than chlorination (Echobichon et al. 1977). Halogen acne may be a more specific term but the definition chloracne will be not abandoned. Chloracne, or something very similar clinically and histologically, has been produced in only three experimental animals: the rhesus monkey, the hairless mouse, and most important, the rabbit (inner surface of the ear).

Cutaneous carcinogenesis.

Research in the field of malignant transformation of skin cells remains in its infancy. However, there is now clear evidence that enzyme-mediated transformation of chemicals and carcinogens (e.g., benzo (a)pyrene) into reactive metabolites such as bay region diol epoxides can mediate toxic effects by binding to macromolecules, thereby initiating the formation of cancer. The role these enzymes play in toxic responses in human skin remains speculative.

One of the major advances in this field has been the availability of systems for culturing mammalian keratinocytes. These culture systems provide an useful approach to the monitoring of effects of environmental chemicals on a variety of biologic and biochemical systems. Whereas cultured keratinocytes have been exploited for a large number of studies of the factors that regulate proliferation and for studies related to chemical carcinogenesis, they have not been sufficiently well utilized to assess the toxic effects of environmental chemicals (Yuspa 1981).

Recent studies have shown that tumor promoters such as the phorbol esters can influence the activity of enzymes such as ornithine decarboxylase, which can participate in the regulation of proliferation. In addition, it is now clear that these compounds can influence cell membrane receptors; for example, the receptor for epidermal growth factor, as well as various proteases, cells of which may be important in mediating their effects

(Slaga et al. 1980).

Studies of mechanisms of tumor promotion, particularly, indicate that several semi-independent steps, affected differentially by various environmental chemicals, can be involved.

Discuss examples of dermal toxicity of pesticides.

The transition from natural to synthetic pesticides occurred during and afer World War II. Soon after DDT (1942) a large number of additional chlorinated hydrocarbons appeared. Allergic contact dermatitis due to DDT has never been convincingly reported (Niedelman 1946).

The organic phosphates have been specially attractive, not only because of their effectiveness but because they rapidly degrade in the environment. Most organic phosphates produce little or no primary irritation. Experimentally parathion and malathion have been reported to be potent contact sensitizers (Parmintieri 1960) but considering their extensive use, dermatitis is rare.

Concerning to the dermatitis from malathion it was shown that the allergen was not the chemical itself but diethylfumarate present as contaminant (Milby et al. 1964).

In an epidemiologic study of 216 patients with contact dermatitis from pesticides in Japan between 1968 and 1970, 18.5% of the cases were due to organophosphates (Matsushita et al. 1980). In a more recent study the Author analyzed 202 patients with contact dermatitis from organophosphorus insecticides: from the case analysis the solitary compounds of organophosphorus insecticides attributed to the dermatitis were DDVP, salithion, sumithion, phosvel and malathion in order (Matsushita et al. 1985) (Table 6). The regions of the dermatitis were finger (62%), face (40%), forearm (32%) neck and nape (30%).

About one quarters of the cases had complication with symptoms of acute poisoning from organophosphorus compounds.

160

Table 6. Causative organophosphorus insecticides of dermatitis.

Pesticide	Solitary	Numerous
DDVP	11	55
Salithion	4	37
Phosvel	4	23
Sumithion	3	24
Malathion	3	15
Diazinon	2	16
Kitazin P	2	3
Cyanox	0	22
Kilvan	0	20
Disyston	0	4
Supracide	0	3
Others	21	70

"Solitary" and "numerous" mean that the
dermatitis was estimated to be occurred by
a solitary compound and numerous compounds
in use, respectively. From Matsushita
(1985) (33).

Fungicides

Fungal diseases of plants have had an enormous impact on history. The great
potato famine in Ireland (1845-49) due to a potato fungus, caused the great
migration to the United States. About 40 fungicides are used on crops today
(Adams 1983).

In 1967 Fregert (Fregert 1967) reported allergic contact dermatitis in a fruit farmer from contact with captan (N-trichloromethylmercapto-Δ^4-tetra-hydrophthalimide). Marzulli and Maibach demonstrated that captan in a concentration of 1% is a significant contact allergic sensitizer (Marzulli et al. 1973).

Among the class of fungicides chloropicrin (trichloronitromethane) and dichlone (2,3-dichloro-1,4-naphthoquinone) are strong irritant, while dermatitis are demonstrated for benomyl, thiram, maneb and zineb (dithio-carbamate).

In 1975 Camarasa in Barcelona reported 4 of 41 workers in a difolatan packaging area with severe eczematous dermatitis and urticaria (Camarasa et al. 1975). Other studies (Cottel 1972) confirmed allergic contact dermatitis due to this fungicide.

Herbicides

An herbicide is any substance used to kill plants or to interrupt their normal growth. The most rapid development of the production was occurred since 1945 with the introduction of a large number of selective organic herbicides as phenylureas, phenylcarbamates, acylanilides, chlorinated aliphatic acids, thiocarbamates and the s-triazine herbicides.

An example of skin toxicity is paraquat. A study by Howard (1979) of 36 workers employed in the manufacture of paraquat revealed several workers with an acute irritant bullous dermatitis and 3 with finger nail damage but neither chronic skin effects nor evidence of allergic sensitization was detected.

Some other components of this class as propachlor, randox, dazomet, lasso showed irritating properties resulting in contact dermatitis.

Contamination of 2,4,5-T with TCDD.

2,4D (2,4-dichlorophenoxyacetic acid) and 2,4,5T (2,4,5 trichlorophenoxy-acetic acid) are selective herbicides and among the first synthetic herbi-cides used.

A 50:50 mixture of their butyl esters is Agent Orange, extensively used as a defoliant. Agent Orange was contaminated with dioxin (TCDD), a powerful chlorinacneigen and possible teratogen, as well as liver toxin (Taylor 1979).

Chloracne is defined as an acneiform eruption due to poisoning by halogenated aromatic compounds having a specific molecular shape (Poland et al. 1977).

Table 7. Chloracnegens in humans

Chloronaphthalenes (CN$_s$)
Polychlorinated biphenyls (PCB$_s$)
Polybrominated biphenyls (PBB$_s$)
Polychlorinated dibenzofurans (PCDF$_s$)
Polychlorinated dibenzodioxins (PCDD$_s$)
Tetrachloroazobenzene (TCAB)
Tetrachloroazoxybenzene (TCAOB)

In table 7 are reported the classes of substances clearly proved to have caused chloracne in humans. During the accident of Seveso on July 10, 1976 mainly 2,4,5 trichlorophenate was discharged in the air. Who were caught in the cloud whit sweasty skin or who handled objects covered with the discharge, suffered chemical burns. As the burns faded, leaving hyperpigmentation, the chloracne developed. There were 8 severely and 11 moderately burned children with a curious type of chloracne which, the more severe cases, formed diffuse thick hyperkeratotic sheets as every follicle had formed a small keratinous cyst.

There were 168 other cases of chloracne all mild to minimal in extent.

Unlike the children, fewer than 5% of the adults burned developed chloracne and this was of a very minor degree.

Probably the major contact at Seveso was external and this accident showed that chloracne is the most sensitive marker of human TCDD poisoning.

Together with clinical disorders in noncutaneous systems, TCDD also produces porphyria in experimental animals (Rose et al. 1976).

Arsenical pesticides.

Arsenic is an ingredient of many insecticides and fungicides and is one of the principal ingredients of agricultural sprays and powders. For many years arsenic has been the most important single cause of accidental death associated with pesticides in the USA.

In this country an estimated 50-60 million pounds of arsenic trioxide are used annually.

The 1975 NIOSH criteria document on arsenic estimates that 1.5 million U.S. workers are exposed in occupations such as copper smelting and other additional used.

In a study of arsenic trioxide exposure of 348 men dermatitis was found to be the most common lesion. Ulceration of the nasal septum, irritation of the nasal and pharyngeal mucosa and conjuctivitis are the next most frequent lesions (Pinto et al. 1953).

Detailed epidemiologic studies in Taiwan have shown that in aereas where drinking water has high arsenic content the prevalence of skin cancer is 25-30 times that of Western countries and 7 times greater than that of general population of Taiwan (Yeh 1963).

Many observers have reported that exposure to arsenic causes dermatitis, melanosis, epidermal hyperplasia, particularly in the form of keratoses of the palms and soles, although they may occurr elsewhere on the body. While these are noninvasive, they are considered to be premalignant and in a small number of cases can progress to squamous cell carcinomas (Graham et al. 1966).

There is no doubt that compounds similar or identical with those used as pesticides have caused skin cancer in man. Squamous cell carcinoma has been associated with arsenic exposure. On the skin, these carcinomas tend to be multifocal and to develop from a premalignant base.

Arsenic seems to inhibit "dark repair enzymes" which repair DNA damage caused by U.V. irradiation (Jung et al. 1969).

Most studies of workers in copper smelting and arsenical pesticide production do not evaluate worker incidence of skin cancer. Other than Paris's study (1920), skin cancer incidence was increased in workers involved in sheep dip manufacture (sodium arsenite) (Hill et al. 1948).

A different case: paresthesia due to exposure to pyrethroid insecticides.

The first synthetic pirethroid was developed in the late 1940s; however, it was not until 1973 that the first photostable pyrethroid was synthesized. In comparison with commonly applied broad-spectrum insecticides such as organophosphates and carbamates, synthetic pyrethroids have a high insecticidal activity, a low toxic effect in mammals, are rapidly metabolized, and leave virtually no residue in the biosphere. The increased use of pyrethroid insecticides during recent years has established them as a major class of pesticide chemicals.

Fenvalerate is a highly active phenyl acetate ester of pyrethroid alcohols: the dermal LD_{50} in the rabbit is greater than 2,500 mg/kg.

As the field use of fenvalerate has increased in recent years, several reports of cutaneous paresthesias among occupationally exposed individuals have appeared. These cutaneous sensations topically occur without erythema, edema, vesiculation, or any other sign of cutaneous irritation.

The effect have been described as transient tingling itching sensations, most frequently experienced on facial skin developing with a latent period of 10-30 minutes to three hours and persisting for about 30 minutes to eight hours and were rarely present the next morning (LeQuesne et al. 1980, Kolmodin-Hedman 1982, Knox et al. 1984).

Vitamin E was found to be (fig. 2) a highly efficacious therapeutic agent for synthetic pyrethroid exposure. It is a well known biological antioxidant that inhibits peroxides from accumulating and protects cells from the scavenging effects of free radicals and also ensures the stability and integrity of biological membranes (Flannigan et al. 1985).

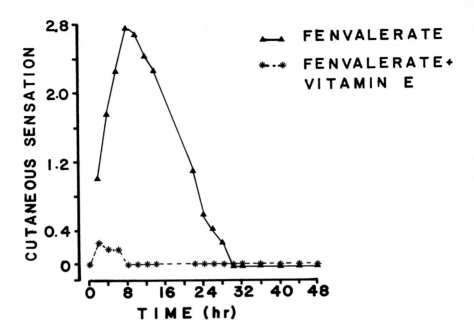

Fig. Therapeutic inhibition of formulated fenvalerate (0.13 mg/cm^2) with vitamin E acetate (1 mg= 1 IU) on people. (From Flanningan et al. 1985).

Conclusion

The development of a complex industrial society has increased the number of physical factors with which man must deal. Man has a added greatly to the number of noxious, irritating and potentially dangerous agents in his environment.

The manmade changes in the environment are occurring very rapidly. For example, by affecting the ozone layer, by products from fertilizer, spray cans, and supersonic transport may, in a few decades, change the spectral distribution of sunlight, which took many millions of years to evolve.

Many of these chemical and physical agents can irritate the skin. These substances cause an inflammatory reaction - a primitive defensive reaction which dilutes, buffers, or otherwise gets rid of offending agents.

Research on the barrier function of skin and how pesticides penetrate it has been very meager. Systematic development of new models for the study of percutaneous absorption should be encouraged. Some aspects that need consideration are how penetration is affected by hydration, by the action of specific organic solvents, and by the molecular structure and physical properties of the penetrant. With this information, new methods could be developed to avoid the penetration of pesticides into the skin. Studies should be undertaken to learn more about the cellular and biochemical characteristics of nonallergic inflammatory reactions and to determine how the molecular characteristics of the irritant produce different pathological patterns. Efforts should be focused on ubiquitous marginal irritants. There is a need to correlate molecular structure and irritant capacity and to examine the action of marginal irritants on cell organelles.

Studies should be developed to determine what cellular and biochemical factors participate in reversing the inflammatory response upon continued exposure. The protein, carbohydrate, and lipid components, enzyme activity, lysosomal components, and kinens of normal, irritated, and postinflamed skin should be examined.

We need to carefully appraise current methods and, more important, develop new methods for predicting the cutaneous hazards of environmental agents. For example, there is a need to develop better methods to detect ubiquitous marginal irritants, acnegenic substances, chemicals that produce granulomas, and substances or processes that alter skin pigment. Animal and human skin, as well as cells in culture, are readily accessible models for use in developmental studies.

REFERENCES

Adams R.M (1983). Pesticides and other agricultural chemicals. In: "Occupational skin disease". Grune and Stratton eds., New York p. 316-378.

Baginsky E. (1978). Occupational disease in California. San Francisco 1982. Division of Labor. Statistics and Research. California Department of Industrial Relations.

Bainova A. and Tomova L. (1980). Skin sensitization after occupational contact with pesticides. Gigienya i Zdraveopazvane 23, 361-367.

Bickers D.R., Kappas A. (1978). Human skin aryl hydrocarbon hydrolase. Induction by coal tar. J. Clin. Invest., 62, 1061-68.

Bradway D.E., Shafik T.M. and Lores E.M. (1977). Comparison of cholinesterase activity, residue levels, and urinary metabolite excretion of rats exposed to organophosphorus pesticides. J. Agric. Food Chem. 25, 1353-1358.

Camarasa G. (1975). Difolatan dermatitis. Contact Derm. 1, 127.

Cottel W. (1972). Difolatan Contact Dermat. News 6, 252.

Durham W.F. and Wolfe H.R. (1962). Measurement of the exposure of workers to pesticides. Bull. W.H.O. 26, 75-91.

Durham W.F., Wolfe H.R. and Elliot J.W. (1972). Absorption and excretion of parathion by spraymen. Arch. Environ. Health. 24, 381-387.

Echobichon D.J., Hansell M.M., Safe S. (1977). Halogen substituents at the 4 and 4' positions of bihenyl: influence on hepatic function in the rat. Toxicol. Appl. Pharmacol. 42, 359-366.

Feldman R.J., Maibach H.I. (1974). Percutaneous penetration of some pesticides and herbicides in man. Toxicol. Appl. Pharmacol. 28, 126-132.

Flannigan S.A., Tucker S.P., Key M.M., Ross C.E., Fairchild E.J., Grimes B.A. and Harrist R.B. (1985). Synthetic pyrethroid insecticides: a dermatological evaluation. Br. J. Ind. Medic. 42, 363-372.

Fouts J.R. (1982). The metabolism of xenobiotics by isolated pulmonary and skin cells. Trends Pharmacol. Sci. 3, 164-166.

Franklin C.A., Fenske R.A., Greenhalgh R., Mathieu L., Denley H.V., Leffingwell J.T. and Spear R.C. (1981). Correlation of urinary pesticides metabolite excretion with estimated dermal contact in the course of occupational exposure to Guthion. J. Toxicol. Environ. Health 7, 715-731.

Franklin C.A., Greenhalgh R. and Maibach H.I. (1983). Correlation of urinary dialkyl phosphate metabolite levels with dermal exposure to azinphos-methyl. In: "Human welfare and the environment". Miyamoto J. and Kearney P.C. eds. Pergamon Press, Oxford, pp. 221-226.

Franklin C.A. (1984). Estimation of dermal exposure to pesticides and its use in risk assessment. Can J. Physiol. Pharmacol. 62, 1037-1039.

Fregert S. (1967). Allergic contact dermatitis from the pesticides Captan and Phaltan. Contact. Derm. Newsl. 2, 28.

Gaines T.B. (1960). The acute toxicity of pesticides to rats. Toxicol. Appl. Pharmacol. 2, 88-99.

Gellin G.A., Maibach H.I., Misiaszek M.H., Ring M. (1979). Detection of environmental depigmenting substances. Contact Dermatitis 5, 201-213.

Graham J.H. and Helwig E.B. (1966). Cutaneous premalignant lesions. In "Advances in biology of skin" vol. 7 Carcinogenesis (Montagna M., Dobson R.L. eds.) P. 227, Pergamon, N.Y.

Hill A.B., Faning E.L. (1948). Studies in the incidence of cancer in a factory handling inorganic compounds of arsenic. Br. J. Med. 5, 1.

Howard J.K. (1979). A clinical survey of paraquat formulation workers. Br. J. Ind. Med. 36, 220-223.

Hudson R.H. et al. (1979). Acute oral and percutaneous toxicity of pesticides to mallards: correlation with mammalian toxicity data. Toxicol. Appl. Pharmacol. 47, 451-460.

Knox J.M. II, Tucker S.B., Flannigan S.A., (1984). Paresthesia from cuta-
    neous exposure to a synthetic pyrethroid insecticide. Arch. Dermatol.
    120, 744-746.
Kolmodin-Hedman B., Swensson A., Akerblan M. (1982). Occupational expo-
    sure to some synthetic pyrethroids (permethrin and fenvalerate). Arch.
    Toxicol. 50, 27-33.
Kundiev J.I. (1975). Pesticide absorption through the skin and prophy-
    laxis of the intoxications. Kiev. Zrorovia, pp. 1-199.
Jung E.G., Trachesel B., Immich. H. (1969). Arsenic as an inhibitor of the
    enzymes concerned in cellular recovery (dark repair). German Med. Mth
    14, 614-616.
LeQuesne P.M., Maxwell I.C., Butterworth S.T. (1980). Transient facial
    sensory symptoms following exposure to synthetic pyrethroids: A
    clinical and electrophysiological assessment. Neurotoxicology 2, 1-11.
Maibach H.I., Feldman R.J., Milby T.H. and Serat W.F. (1971). Regional
    variation in percutaneous penetration in man. Archives of Environ. He-
    alth, 23, 208-211.
Maibach H.I. and Feldmann R.J. (1974). Systemic absorption of pesticides
    through the skin of man. In "Occupational Exposure to Pesticides:
    Report to the Federal Working group on Pest Management from the Task
    Group on Occupational Exposure to Pesticides, Appendix B, pp. 120-127.
U.S. Government Printing Office 1975, O-551-026, Washington.
Marzulli F.N., Maibach H.I. (1973). Antimicrobials: Experimental contact
    sensitization in man. J. Soc. Cosmet. Chemists 24, 399-421.
Matsushita T., Nomura S., Waktsuki T. (1980). Epidemiology of contact
    dermatitis from pesticides in Japan. Contact Dermatitis 6, 255-259.
Matsushita T., Aoyama K., Yoshimi K., Fujita Y., Ueda A. (1985). Allergic
    contact dermatitis from organophosphorus insecticides. Industrial
    Health 23, 145-153.
Milby Th. H., Epstein W.L. (1964). Allergic contact sensitivity to Mala-
    thion Arch. Environm. Health 9, 434-437.
Morison W.L., Parrish J.A., Woehler M.E. et al. (1981). The influence
    of ultraviolet radiation on allergic contact dermatitis in the guinea
    pig. I. U.V.B. radiation Br. J. Dermatol 104, 161-164.
Niedelman M.L. (1946). Contact dermatitis due to DDT. Occup. Med. 1,
    391-395.
O.S.H.A., Standards advisory Committee on cutaneous hazards. Report to
    Assistant Secretory of Labor. (1978). December 19.
Pannatier A., Jenner P., Testa B. and Etter J.C. (1978). The skin as a drug
    metabolizing organ Drug Metab. Rev. 8, 319-343.
Parkinson E.K., Newbold R.F. (1980). Benzo(a) pyrene metabolism and DNA
    adduct formation in serially cultivated strains of human epidermal
    keratinocytes. Int. J. Cancer 26, 289-299.
Paris J.A. (1920). Pharmacologici. 3rd ed., p. 282 Phillips, London.
Parmintieri G. (1960). On the experimental sensitization with Parathion
    Ital. Rev. Derm. 7, 48-52.
Pinto S.S., McGill C.M. (1953). Arsenic trioxide exposure in industry
    Ind. Med. Surgery, July 281-287.
Poland A., Glover E. (1977). Chlorinated byphenyl induction of aryl hy-

drocarbon hydroxylase activity; a study of the structure activity relationships. Mol. Pharmacol. 13, 924-938.

Rose J.Q., Ramscy J.C., Wertzler T.H., Hummel R.A. and Gehring P.J. (1976). The fate of 2,3,7,8-tetrachlorodibenzo-p-dioxin following single and repeated oral doses to the rat. Toxicol. Appl. Pharmacol. 36, 209 220.

Skafik T., Bradway D.E., Enos H.F. and Yobs A.R. (1973). Human exposure to organophosphorus pesticides. A modified procedure for the gas liquid chromatographic analysis of alkyl phosphate metabolites in urine. J. Agric. Food Chem. 21, 625-629.

Slaga T.J., Fisher S.M., Nelson K. et al (1980). Studies on the mechanism of skin tumor promotion: Evidence for several stages in promotion. Proc. Natl. Acad. Sci. USA 77, 3659-3669.

Stobo J.D., Kennedy M.S., Goldyne M.E. (1979). Prostaglandin E_2 modulation of the mytogenic response of human T. cells J. Clin. Invest. 64, 1188-1195.

Taylor J.S. (1979). Environmental chloracne: update and overview. Ann. NY. Acad. Sci. 320, 295-307.

U.S. Department of Labor. Bureau of Labor. Statistics, occupational injuries and illnesses in the United States by Industry 1980. April, (1982) Bulletin 2130.

Wester R.C and Maibach H.I. (1983). Cutaneous pharmacokinetics: 10 steps to percutaneous absorption. Drug Metab. Rev. 14, 169-205.

WHO Chronicle (1975). 29, 397-401.

Wolfe H.R., Durham W.F. and Armstrong J.F. (1967). Exposure of workers to pesticides, 14, 622-633.

Wolfe H.R., Armstrong J.F., Staiff D.C. and Comer S.W. (1972). Exposure of spraymen to pesticides. Arch. Environ. Health, 25, 29-31.

Yeh S. (1963). Relative incidence of skin cancer in Chinese in Taiwan with special reference to arsenical cancer. Natl. Cancer. Inst. Monogr. 10, 81-102.

Yonemoto K., Gellin G.A., Epstein W.L. et al. (1983). Reduction in eumelanin by the activation of glutathione reductase and gamm glutamyl transpectidase after, exposure to a depigmenting chemical. Biochem. Pharmacol. 32, 1379-82.

Yuspa S.H. (1981). Chemical carcinogenesis related to the skin. Parts I and II, In "Progress in Dermatology". Vol. 15, No. 4, and Vol. 16, No. 1, Evanston, IL, 1982, Dermatology Foundation.

Environmental Distribution and Fate of Pesticides. A Predictive Approach

D. Calamari
Institute of Agricultural Entomology
University of Milan
Via Celoria 2
20133 Milano
Italy

E. Bacci
Department of Environmental Biology
University of Siena
Via delle Cerchia 3
53100 Siena
Italy

Introduction

Since World War II, due to technological growth, large quantities of synthetic chemicals, probably unknown to the biota before (xenobiotics), have been
introduced into the natural systems. Some of these compounds are quickly degraded or transformed into natural or inert substances, others are long living and easily transported far from the immission areas.
Around the sixties the scientific community became aware of the potential
danger that these xenobiotics could provoke to the ecosystems and the number
of scientific papers on environmental problems increased exponentially.
For many years, in fact, the information on the distribution and fate of che
mical substances, after their use in domestic, industrial and agricultural
activities was largely gained in retrospect from empirical observations, after wide monitoring campaigns with an enormous number of chemical analysis.
The retrospective approach left wide margins for errors in the environmental
management of organic chemicals and the possibility for great scale undesira
ble effects remained largely uncontrolled.
However, in the early seventies both the scientific community and administra
tors started to think in predictive terms.
Several investigators advanced approaches to predict the behaviour of new
chemicals by comparing laboratory measured properties with those of compounds,
such as some chlorinated hydrocarbons, for which more environmental data were
existing.
The concept of environmental chemodynamic was then diffused as a holistic ap

NATO ASI Series, Vol. H13
Toxicology of Pesticides: Experimental, Clinical
and Regulatory Aspects. Edited by L. G. Costa et al.
© Springer-Verlag Berlin Heidelberg 1987

proach toward chemical environmental behaviour (Haque and Freed, 1975), and in following years many other publications appeared on the same subject (Bau ghman and Lassiter, 1978; Haque, 1980; Neely, 1980; Hutzinger, 1980, 1982; Gunther, 1983; Neely and Blau, 1985; Sheehan et al., 1985; GSF, 1985).

The predictive approach received an important and essential impulse in USA from the approval of the Toxic Substances Control Act (1978) and in Europe from the Directive and Dangerous Substances (1979). Both these regulations call for a certain amount of information before a chemical substance being introduced into the market.

According to these approaches the Organization for Economic Cooperation and Development started a project on hazard assessment (OECD, 1982). This proce dure is basically a comparison between two terms: exposure versus effects. Bro-Rasmussen and Christiansen (1984) described per extenso this type of eva luation. One side of a hazard assessment procedure is therefore how to calcu late the potential environmental exposure.

In this paper the possibility of predicting environmental behaviour of pesticides by means of measurements which could be made in laboratory or by means of limited field tests will be discussed.

Partition analysis

Pesticides are introduced into the environment for specific purposes and in defined ways, however these substances will move, as any other molecule, from their point of entry to their final destination i.e. the environmental compartment for which they have more affinity. From this last, if it is not a "sink" place but a "reservoir", the chemicals can be transferred again towards other compartments.

In the mean time pesticides can undergo chemical transformations in every environmental compartment including biota.

Figure 1 shows in a schematic way the major environmental compartments and the modality of transport among them. For each compartment the relevant degradation processes are also listed.

To evaluate the environmental distribution of a pesticide the parameters are: Henry constant (H), water solubility (S), soil sorption coefficient (Koc) and n-octanol/water partition coefficient (Kow). The numerical value of each parameter indicates the degree of affinity for the four basic ecological compartments: air, water, soil and biota.

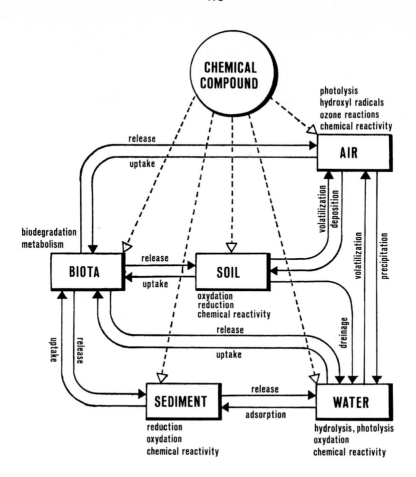

Fig. 1 - Diagram showing transport and transformation processes for chemical compounds in environmental compartments.

Physico-chemical properties

The physico-chemical properties from which the above reported partitioning properties are calculated are: molecular mass, melting point, vapour pressure and water solubility.

Water solubility (S) and vapour pressure (v.p.), are essential in determining the environmental distribution of a given substance: with a high S the chemical will show considerable affinity for the aqueous phase, while a low S value indicates a tendency to escape from the water. S also regulates uptake by biota and adsorption-desorption in water media on solid substrates. Vapour pressure indicates the tendency of a chemical to leave, as vapour, its pure phase; volatilization phenomena are strictly related to this property. Melting point (m.p.) indicates the interaction energy in crystals, which opposes solvent actions.

Partition coefficients

The Henry constant indicates the equilibrium partitioning between air and water and can be calculated as $H = v.p./S$.

The adsorption processes in soils, sediments and particulate matters in aqueous solutions have been described according to the Freundlich adsorption isotherm, which could be fitted for many diluted solutions by the following equation:

$$x/m = K \, C^{1/n}$$

where x/m is the amount of adsorbate per unit of adsorbent, C is the equilibrium concentration of adsorbate, K and n are constants related to the bonding energy.

At low pollutant concentrations the sorption isotherm onto soils and sediments are linear and reversible

$$S = K_p \, C$$

where S is the concentration of the chemical in the adsorbed phase, C the concentration of the chemical in the water phase, and K_p the partition coefficient.

Different soils and sediments, normalised to the organic carbon (oc) can show very similar K_p values, the adsorption being mostly on organic materials. If so the previous relation between S and C becomes:

$$S = K_{oc} \, C$$

where K_{oc} is the organic carbon sorption coefficient, related to K_p as follows:

$$K_{oc} = K_p/F_{oc}$$

where F_{oc} represents the organic carbon fraction in the soil and sediment phase.

The dimensionless K_{oc} gives therefore a measure of the affinity of a molecule for a soil.

The n-octanol/water partition coefficient (Kow) represents the ratio between the concentration in n-octanol phase and in water phase at the equilibrium and it is an estimation of the hydrophobicity or lipoaffinity of a substance dissolved in water. From Kow an estimate of the bioconcentration factor

(BCF) can be obtained assuming a first order or a pseudo-first order kinetic and a linear two compartment model.

All these parameters including basic physico-chemical characteristics can be found in the scientific literature or obtained by means of laboratory measurements. They can also be calculated by means of property-property correlations or with the fragment constant methods or by means of topological indices.

Looking at table 1, the concept outlined above found practical evidence where a classification scheme for pesticides is proposed. In fact if a compound has for example a solubility at $g.1^{-1}$ level it has a high affinity for the water compartment, if the Kow is lower than 10^3 the substance does not accumulate in biota etc.

Table 1 - Classes of affinity of pesticides for the different environmental compartments in relation to the physico-chemical characteristics of the molecules.

Affinity	Air	Water	Soil	Biota
	H	S	Koc	Kow
	$Pa.m^3.mol^{-1}$	$g.1^{-1}$		
Low	$< 10^{-3}$	$< 10^{-3}$	< 1	$< 10^3$
Medium	$10^{-3}-1$	$10^{-3}-1$	$1-10^3$	10^3-10^5
High	> 1	> 1	$> 10^3$	$> 10^5$

One can do even better classification increasing the degrees of affinity (i.e. very low, very high etc.) considering that for example the solubility has a range of about seven order of magnitude and Koc about eleven.

However, this type of classification is valid as a first approach, being particularly useful only for molecules with characteristics at the extreme value levels, moreover the ranking can widely vary if organic industrial chemicals are included and the classes of affinity modified in accordance.

Finally this first approach does not give an idea of the behaviour of the compounds when the properties are considered all together, in an integrated way, better simulating what is happening in the real world.

To overcome these limitations models of compartmental analysis were proposed (Neely, 1979; Mackay, 1979; Frische et al., 1982). The input to these models

being the described physico-chemical parameters and the partition coefficients and the output being the expected percentage distribution in the
main environmental compartments. One of these models will be shortly descri
bed below.

Transformation kinetics analysis

The environmental fate is evaluated on the basis of the persistence of the
pesticide, which can, in natural conditions, be degraded in various ways ac
cording to its molecular structure.
The degradation processes are both biotic and abiotic: the former are biode
gradation and metabolism, the latter mainly photolysis, hydrolysis and oxy-
dation. All these reactions can be assumed as following first order kinetics.
If so, the rate of each degradative process is expressed as the product of
the concentration of the chemical in the compartment considered and the rate
constant (t^{-1}). Consequently, all reaction rates in a given phase can be
added, obtaining a total first order rate constant, K, and by multiplying
this and the concentration of the chemical in the compartment, C, the total
degradation rate of the compartment, KC, can be calculated:

$$\text{degradation rate} = K_1C + K_2C + K_3C + ...+ K_nC = KC$$

As one can easily see, the importance of an environmental compartment as
sink for a given chemical is strictly dependent on its total degradation-
rate constant and on its potential for attaining a high concentration of
the pollutant.

Mass balance

In order to be able to calculate the predicted environmental concentration
together with the partition and transformation kinetic analysis, a mass ba
lance has to be made by knowing the quantity of pesticide applied and the
extension of the area involved.
Moreover the presence and the level of concentration of a chemical substan
ce in a given compartment is not only a function of its possibility to be
degraded (persistence) but also of the transfer rate toward other compartments and of advection.
Advection, which is generally negligible for soils, is particularly impor-

tant for fluid phases (air, water), and can be considered as a first order process with rate constant K_a (as t^{-1}), defined as follows (assuming steady-state conditions):

$$K_a = I/Q$$

where I is the input rate (or output), and Q is the total amount of the chemical in the compartment.

The overall mean residence time of the chemical in the compartment in steady-state conditions, T, will be:

$$T = 1/(K+K_a)$$

where K and K_a are the overall degradation rate constant and the advection rate constant, respectively, previously defined.

Fugacity models

In order to link all the properties and the processes already described a number of evaluative models have been proposed (i.e. the fugacity approach). Fugacity (f) is an old physico-chemical concept that Mackay (1979) rediscussed in new terms and defined as the tendency, for a chemical substance, to escape from one phase to another. This property can be calculated in units of pressure (Pa).

Subsequently Mackay and Paterson (1981) proposed an evaluative model of $1 km^2$, they called "unit of world", divided into six compartments with defined quantities of materials. They also introduced the concept of environmental capacity for each compartment ($Z = mol.m^{-3}.Pa^{-1}$) from which the theoretical concentrations ($C = mol.m^{-3}$) could be calculated after an immission of a given amount of chemical compound

$$C = f Z$$

The equilibrium being attained when the fugacities are equal in all the compartments

$$f_1 = f_2$$

therefore

$$C_1/Z_1 = C_2/Z_2$$

and

$$C_1/C_2 = Z_1/Z_2 = K_{12}$$

K_{12} is the partition coefficient determining the distribution of the substance between two phases.

The capacities of each compartment (Z) can be determined, as function of par
tition coefficients. If equilibrium, well mixing, no reaction and no advection is assumed the relative mass distribution and relative concentrations
can be calculated.

In practice after the application of the fugacity model (level I) one can
know in which compartment most of the compound is found and where the highest concentrations in the "unit of world" are.

Level II is also at the equilibrium and it includes reactions of transforma
tion and advection. Kinetics of transformation can be derived from literatu
re and a transformation matrix, which gives the persistence time in a given
environment, can be prepared.

Level III (Mackay and Paterson, 1982) is a more complex steady-state non
equilibrium system which gives an idea of the transport between phases, whi
le level IV includes variations in the emissions during time.

Other types of more complex evaluative models such as EXAMS (Burns et al.,
1981) and QWASI (Mackay et al., 1983) have also been built up and allow fur
ther stages of assessment of the environmental behaviour of chemicals to be
positively performed.

Application to case studies

In order to estimate value and limitations of the predictive approach here
presented, some examples will be discussed in particular through the experience of the research groups to which the authors belong.

Theoretical applications

The first exercise was made on four pesticides (Calamari and Vighi, 1983)
with different chemical characteristics and, in addition to the original
standard procedure, calculations were made considering emissions correspon
ding to common agricultural treatments. By comparing the potential environ
mental concentrations to toxicity data on different organisms a simple hazard assessment was made, including considerations on persistence evaluated
by means of a transformation matrix.

In table 2, 3 and 4 the relevant results of the exercise are reported, for
malathion as an example, repeating the calculations in similar conditions.

Table 2 - Malathion: Calculations of Fugacity

Basic data
    molecular weight   = 330.3
    solubility in water = 145 mg.l^{-1}
    vapour pressure   = 5.3 · 10^{-3} Pa

Calculated data
    Kow   = 4860
    Koc   = 2990
    H   = 1.21 · 10^{-2} Pa m^3 . mol^{-1}
    BCF   = 272
    Kp soil = 59.98
    Kp sed. = 119.97

Data from Mackay's models
    density of soil and sediments = 1500 kg.m^{-3}
    density of biota and water   = 1000 kg.m^{-3}
    density of air   = 1.19 kg.m^{-3}
    organic C content in soil   = 2%
    organic C content in sediments = 4%
    temperature   = 25°C

Table 3 - Partition Analysis: characteristics of environmental compartments and malathion concentrations on 100 moles emission.

Compartment	Volume	Z	Amount	Concentration	
	m^3	mol.m^{-3}.Pa^{-1}	mol	mol.m^{-3}	µg g^{-1}
Air	6 ·10^9	4.04·10^{-4}	0.1967	3.27·10^{-11}	0.09·10^{-6}
Water	7 ·10^6	8.28·10^1	47.0889	6.73·10^{-6}	1.33·10^{-1}
Soil	4.5·10^4	0.75·10^4	27.2375	6.05·10^{-4}	2.22·10^{-3}
Biomass	7	2.25·10^4	0.0128	1.83·10^{-3}	6.03·10^{-1}
Suspended solids	35	1.49·10^4	0.0424	1.21·10^{-3}	2.66·10^{-1}
Sediments	2.1·10^4	1.49·10^4	25.4217	1.21·10^{-3}	2.66·10^{-1}
		Total amount	100.0000		

Fugacity 8.1215 · 10^{-8} Pa

The partition analysis shows soil sediments and water as the relevant compartments for distribution. The concentrations in water are nearly those indicated as toxic at an acute level (2-20 μg/l as 48 LC50). On the contrary levels in soil and biota are far from those considered potentially dangerous.
However, from table 4, one can see that the water is the compartment where most of the degradation occurs and the half time is relatively short (136 h).

Table 4 - Transformation kinetics analysis for malathion.

Compartment	K h^{-1}	KC $mol \cdot m^{-3} \cdot h^{-1}$	VKC $mol \cdot h^{-1}$	%
Air	-	-	-	-
Water	0.010^b	$6.72\ 10^{-8}$	$4.70\ 10^{-1}$	18.55
Soil	0.005^a	$3.02\ 10^{-6}$	$1.36\ 10^{-1}$	64.09
Biomass	0.014^c	$2.55\ 10^{-5}$	$1.79\ 10^{-4}$	0.02
Suspended solids	0.005^a	$6.05\ 10^{-6}$	$2.11\ 10^{-4}$	0.03
Sediments	0.005^a	$6.05\ 10^{-6}$	$1.27\ 10^{-1}$	17.41

K = transformation rate   a = biodegradation
KC = reaction rate per volume unit   b = hydrolysis + biodegradation
VKC = reaction rate   c = metabolism

$$Persistence = \frac{\Sigma\ VC}{\Sigma\ VKC} = 136\ h$$

For comparing various molecules and as a first estimate of the behaviour of the chemical compounds the exercise was considered quite useful.
The major limitation found is that the model is of a thermodynamic type and does not give any indication of the kinetics of distribution (i.e. times necessary for joining an equilibrium situation).
Other criticisms are of a parametric type as for example there is no possibility to evaluate the role of terrestrial biomasses (vegetation). It is however theoretically possible to improve these simple models refining the volumes of the compartments and eventually including one or more compartments.
The same exercise was repeated on a wider scale with about 30 insecticides

(Calamari et al., 1985). The major limitations identified were not in this case related to the fugacity model but to the quality and the availability of the basic physico-chemical data. For example, for a few of them data were vague or not available in the reference books (i.e. solubility in water: miscible, vapour pressure less then ..., negligible etc.). For other substances such as carbaryl the vapour pressure was in different authoritative textbooks from $2.64 \cdot 10^{-3}$ Pa to 6.665 Pa while the Pesticide Manual (Worthing and Walker, 1983) indicates $6.65 \cdot 10^{-1}$ Pa; for cypermethrin the water solubility was in a range from 0.010 to 0.200 mg/l etc.

A second source of uncertainty was related to the characteristics calculated by means of property-property equations. In fact, a number of equations are available in the literature but the best correlations have been obtained with homogeneous groups of chemical compound and the choice of the most appropriate equations is sometimes not easy. Moreover the extreme values have the greatest margins of errors.

As a third case the fugacity model was succesfully applied in the preparation of the ecotoxicological profile of xilenes (Jori et al., 1986) with the confirmation of air as the major compartment at risk in long-term and the identification of photolysis in air as the most relevant process in terms of percentage of substance transformed.

Laboratory experiments

Notwithstanding the relevance of cypermethrin a few works have dealt with the environmental distribution and fate of this molecule. In order to identify the major environmental reservoirs, potential sinks and degradation compartments, research was performed by using both evaluative models, applying the fugacity approach, and laboratory simulation chambers (Bacci et al.,1987). The research was executed in three steps. As a first step the potential distribution of cypermethrin in arbitrarily simplified environments was calculated by means of the fugacity approach. In a second step a small greenhouse was selected to evaluate the potential of cypermethrin volatilization from treated soils, and to investigate the possibility of translocation from the soil to the leaves of an higher plant species through the root apparatus. Experiments on the possibility for the molecule to be transferred in the water medium were excluded according to theoretical calculations.

In the third step the degradation kinetics of cypermethrin in different soils under controlled laboratory conditions have also been studied.

The experiments gave the predicted results, in fact, according to its physi
cal and chemical properties,cypermethrin has a strong soil affinity and it
is practically non-mobile toward air and water and does not accumulate in
plants.

Moreover it has been confirmed, according to literature data, that the persistence in soils is not very long; the degradation occurs with a half life
of around 40 days.

The combination of evaluative models and a few ad hoc experiments in a simu
lation chamber seems therefore a promising tool for a first approach to the
assessment of basic aspects of environmental behaviour of pesticides, expe-
cially for "new chemicals".

Field studies

In an extensive study on chlorinated hydrocarbons in plant foliage Gaggi et
al. (1985), among the other considerations, concluded that polluted soils
(and sediments) can be considered as reservoirs for hydrophobic persistent
chlorinated hydrocarbons, more than as sinks and that it was possible to
evaluate air pollution levels through foliage data.

To elucidate the first point, and to explain the relative stability of mean
air contamination levels in background conditions, as well as in "hot spot"
situations, Gaggi et al. (1985) used the fugacity model for potential environmental distribution to provide general information on the behaviour of
HCB, α and γ HCH, DDT and its various metabolites and some PCBs.

From those calculation it had been observed that soil and sediments together
contain from 82.3% α HCH to 99,9% p'p' DDT of the total. The vapour movements from polluted soils are important in the dissipation of these contaminants, air being the second compartment involved in terms of percentage of
molecules present; water has much less content. Therefore polluted air
probably represents, at least for these molecules, the main route of entry
for foliage contamination.

Conclusions

In relation to the debate on the usefulness of scenarios and models for stu
dies on the environmental distribution and fate of organic chemicals (GSF,
1985), one can say that, although far from being fully satisfactory, this
part of the hazard assessment scheme has got an acceptable level of develop

ment. An exponentially increasing number of papers has been published on
that subject in the last years. Following this activity a number of occasional and inherent limitations have been identified in the evaluative models approach and are actually open field for research.

The evaluative models have however been proven relevant in a number of situations and will certainly be a powerful tool in the predictive approach
for the environmental management of chemical compounds and in pollution con
trol activities.

References

Bacci E, Calamari D, Gaggi C, Vighi M (1987) Laboratory experiments for the
    prediction of environmental distribution and fate of cypermethrin. Chemo
    sphere (submitted to)
Baughman GL, Lassiter RR (1978) Prediction of environmental pollutant con-
    centration. In Cairns J Jr, Dickson KL, Maki AW (eds) Estimating the ha-
    zard of chemical substances to aquatic life. American Society for Testing
    and Materials ASTM STP 657: 35-54
Bro-Rasmussen F, Christiansen K (1984) Hazard assessment - A summary of ana-
    lysis and integrated evaluation of exposure and potential effects from
    toxic environmental chemicals. Ecological Modelling 22: 67-84
Burns LA, Cline DM, Lassiter RR (1981) Exposure analysis modelling systems
    (EXAMS): user manual and system documentation U.S. EPA Environmental Re-
    search Laboratory, Athens, Ga
Calamari D, Vighi M (1983) Modello per lo studio della distribuzione e del
    destino ambientale degli antiparassitari. La difesa delle piante 6: 381-
    394
Calamari D, Vighi M, Pestalozza E (1985) Applicazione di un modello per lo
    studio della distribuzione ambientale delle sostanze organiche. S IT E
    Atti 5: 123-125
EEC (1979) Council Directive 79/831. Off Jour Eur Comm L 259/10
Frische R, Esser G, Schönborn W, Klöpffer W (1982) Criteria for assessing
    the environmental behaviour of chemicals: selection and preliminary quan
    tification. Ecotoxicol Environ Saf 6: 283-293
Gaggi C, Bacci E, Calamari D, Fanelli R (1985) Chlorinated hydrocardons in
    plant foliage: an indication of the tropospheric contamination level.
    Chemosphere 14:1673-1686
GSF (1985) Environmental modelling for priority setting among existing che-
    micals. Workshop 11-13 Nov 1985 München-Neuherberg Gesellschaft für Stra
    hlen-und Umweltforshung mbH München
Gunther FA (ed)(1983) Residues of pesticides and other contaminants in the
    total environment. Res Rev 85 Springer-Verlag New York Heidelberg Berlin
Haque R (ed) (1980) Dynamics, exposure and hazard assessment of toxic che-
    micals. Ann Harbor Science Ann Harbor Mich
Haque R, Freed VH (eds) (1975) Environmental dynamics of pesticides. Plenum
    Publishing Co New York
Hutzinger O (ed) (1980) The handbook of environmental chemistry. Reaction
    and processes Vol 2 Part A Springer-Verlag Berlin Heidelberg New York
Hutzinger O (ed) (1982) The handbook of environmental chemistry. Reaction
    and processes Vol 2 Part B Springer-Verlag Berlin Heidelberg New York

Jori A, Calamari D, Di Domenico A, Galli CL, Galli E., Marinovich M, Silano V (1986) Ecotoxicological profile of xilenes. Ecotoxicol   Environ Saf 11: 44-80

Mackay D (1979) Finding fugacity feasible. Environ Sci Technol 13:1218-1223

Mackay D, Paterson S (1981) Calculating fugacity. Environ Sci Technol 15: 1006-1014

Mackay D, Paterson S (1982) Fugacity revisited. Environ Sci Technol 16: 654-660

Mackay D, Joy M, Paterson S (1983) A quantitative water, air, sediment inte raction (QWASI) fugacity model for describing the fate of chemicals in lakes. Chemosphere 12: 981-997.

Neely WB (1979) A preliminary assessment of the environmental exposure to be expected from the addition of a chemical to a simulated aquatic ecosystem. Int J Environ Studies 13: 101-108

Neely WB (1980) Chemicals in the environment. Distribution. Transport. Fate Analysis. Marcel Dekker Inc New York and Basel

Neely WB, Blau GE (eds) (1985) Environmental exposure from chemicals. Vol I CRC Press Inc Boca Raton Florida

OECD (1982) Hazard assessment project. a. Exposure analysis b. Health effects c. Natural environment effects. OECD Paris

Sheehan P, Korte F, Klein W, Bourdeau P (eds) (1985) Appraisal of tests to predict the environmental behaviour of chemicals. SCOPE 25 John Wiley and Sons Chichester New York Birsbane Toronto Singapore

US EPA (1978) Toxic Substances Control Act Fed Reg 43: 4108 Washington DC

Worthing CR, Walker SB (eds) (1983) The Pesticide manual. 7 ed The British Crop Protection Council

Ecotoxicology of pesticides: the laboratory and field evaluation of the environmental hazard presented by new pesticides.*

A.R. Hardy
Ministry of Agriculture, Fisheries and Food
Agricultural Science Service
Worplesdon Laboratory
Tangley Place
Worplesdon
Guildford GU3 3LQ
United Kingdom.

INTRODUCTION

Pesticides have an important role to play in the control of pests, pathogens and weeds in agriculture and of pests in public hygiene situations. In order to reduce the environmental impact of such toxic chemicals, it is necessary to maintain strict regulation of the introduction of new pesticides and changes in the use of existing pesticides, to control their use and to monitor their subsequent impact on the environment.

The environmental risks from a given pesticide active ingredient or its formulated product depend on a range of factors: its toxicity to wildlife, its mobility and persistence in the environment, the formulation, how much is applied, the method and timing of application and its frequency of use. This paper will review the evaluation process for pesticides, combining field assessment with laboratory measurements, which is designed to identify environmental hazard and to demonstrate environmental safety.

International agreement between governments has encouraged the harmonisation of regulatory approach and established the principles of safety evaluation and the criteria that should be considered (FAO 1981, FAO 1982). These rely heavily on methodology developed for example by the Environmental Protection Agency (EPA) in the United States and the Pesticides Safety Precautions Scheme (PSPS) in the United Kingdom. This paper concerns the approach taken in the United Kingdom and the experience of monitoring pesticide impact on the environment for 25 years.

THE PESTICIDES SAFETY PRECAUTIONS SCHEME

The objective of pesticide registration in the UK is to safeguard human beings

NATO ASI Series, Vol. H13
Toxicology of Pesticides: Experimental, Clinical
and Regulatory Aspects. Edited by L. G. Costa et al.
© Springer-Verlag Berlin Heidelberg 1987

livestock, domestic animals, beneficial insects, wildlife and the environment against risks which could arise from the use of pesticide products. The PSPS, established in 1957, is a formally negotiated agreement between the Trade Associations representing industry and the Agricultural and Health Departments of government and has been substantially revised since then (MAFF 1986). Most recently this formal agreement has been replaced by legislation. Risk evaluation of a new pesticide involves, the assessment of safety to users and bystanders, to the consumers of food and to the environment. Data are derived from toxicological testing, the measurement of residues in crops and food and from specific environmental tests. Assessment combines the results of laboratory and field tests and allows the stepwise introduction of new pesticides onto the market as information becomes available to provide the necessary assurance that the proposed use is safe. A central principle of registration in the UK is flexibility in identifying the total data requirements. According to the properties and intended use pattern of the particular pesticide, scientific judgements are made in the progressive planning of a test programme rather than initiating a routine and potentially irrelevant series of tests without frequent reassessment (Bunyan and Stanley 1979). Laboratory tests form the first phase of the evaluation process.

LABORATORY DATA REQUIREMENTS FOR ENVIRONMENTAL IMPACT ASSESSMENT

1. Chemical and medical toxicology tests

Data of considerable relevance can be obtained from the results of physical, chemical and medical toxicology laboratory tests which are required for all pesticides. The intrinsic physical and chemical properties of the active ingredient are readily determined in well-established test methods (eg OECD 1981, FAO 1981). Such properties include melting and boiling points, specific gravity and volatility as well as solubility in water and organic solvents. The ease with which a compound degrades for example by oxidation or hydrolysis is also relevant.

Basic toxicology tests on laboratory mammals including rats, mice, rabbits and guinea-pigs, which form part of the primary toxicology requirements for every pesticide, also provide data which can be used to assist environmental assessment. These include acute toxicity, short-term dietary, longer-term chronic and reproductive, mode of action and metabolic studies together with neurotoxicity studies (in the hen).

2. Laboratory tests specifically for environmental hazard assessment

Additional data can be obtained from laboratory tests specifically designed to indicate hazard to wildlife species. Clearly it is not possible to test individually all non-target species that are likely to be exposed to the pesticide in field use but a representative selection of species can be used to indicate the need for field assessment. Such laboratory tests include acute and dietary toxicity studies with two species of birds to include effects on reproduction. Mallard (Anas platyrhynchos), ring-necked pheasant (Phasianus colchicus), feral pigeon (Columba livia), Japanese quail (Coturnux japonica) and northern bobwhite quail (Colinus virginianus) are the species most frequently tested. To identify hazard to aquatic life, chronic toxicity is measured for two freshwater fishes preferably under continuous flow conditions to maintain pesticide exposure. Fish are usually selected from the harlequin (Rasbora heteromorpha), rainbow trout (Salmo gairdneri) or the zebra danio (Brachydanio rerio). The water flea (Daphnia magna is tested as a representative aquatic invertebrate which can also be an important food for freshwater fish. Rainbow trout and catfish (Ictalurus punctatus) may also be used in studies to measure the rate of accumulation of residues during extended exposure to the pesticide.

All pesticide products which will be used in agriculture, horticulture and forestry at a time when bees are at risk, are tested against the honey bee (Apis mellifera) as an example of a beneficial insect important for its pollination activities and honey production. In the laboratory bees may be fed with the pesticide or may receive individual topical applications from a microapplicator (Stevenson 1968). The selection of other beneficial insect species is currently under consideration though the interpretation of the results in relation to field use are not yet clear. Suitable laboratory and field test methods have been developed through the International Organisation for Biological Control of Noxious Animals and Plants (IOBC) for a range of species (Hassan et al 1985). These include, for example, spiders, aphid parasites and carabid beetles. Earthworms are usually considered to be beneficial indicators of soil fertility. Controlled exposure tests have been developed confining worms in containers of standard soil to which the pesticide is applied (PSPS 1981). Mobility of the compound and its behaviour in different soil types can also be measured under controlled laboratory conditions to provide comparative data on the likely behaviour in the field (EPA 1985). In the past test data on the effects of the pesticide on soil function have also been requested but difficulties of interpreting the results limit the value of such studies (Somerville et al 1986).

According to the type of pesticide and its mode of action, other laboratory tests may be useful in predicting likely effects in the field. For example, rodenticides may present a hazard to predatory or scavenging species through secondary poisoning. It is therefore advisable to obtain some data on the risks before field assessment. Test methods exist involving the feeding of contaminated mouse tissue to laboratory rats, or rat tissue to dogs. More specific tests have been developed using the laughing gull (Larus atricilla) (Fink and Jaber 1981), the tawny owl (Strix aluco) (Townsend et al 1981) and the weasel (Mustela nivalis) (Townsend et al 1983). A mammal toxicity test has been developed in the USA using domestic ferrets (Mustela furo) or mink (M. vison) to provide an alternative to the reliance on medical toxicology tests for mammalian data at this stage of testing (Hornshaw et al 1986).

The limited range of laboratory tests outlined above give an indication of the susceptibility of the individual species tested but do not allow realistic prediction of field effects. Susceptibility and toxicity can vary markedly between species (Schafer 1972, Stanley and Bunyan 1979) further limiting the extent to which laboratory results can be extrapolated to other species. A number of reviews of comparative toxicity data have been published from tests with a range of species (Balcomb et al 1984, Hill et al 1975, Hill and Carmadese 1986, Hoffman and Albers 1984, Hudson et al 1984, Schafer 1972, Schafer et al 1983, Tucker and Haegele 1971, Tucker and Leitzke 1979).

Considering the intrinsic properties of the pesticide active ingredient and its formulations measured under laboratory conditions together with a knowledge of the intended use pattern of the chemical, it is possible to predict those situations, eg. crop and time of year, when wildlife may be at risk and where field assessment is necessary. This second phase of environmental testing is therefore very much tailored to the individual pesticide and this is where flexibility is essential (Stanley and Hardy 1983). Though some harmonisation of approach is possible, standardisation of field trial protocols is not. Assessment in the field must be directly relevant to the specific use of an individual product.

FIELD TEST METHODS TO ASSESS THE ENVIRONMENTAL HAZARDS PRESENTED BY PESTICIDES

Historically the approach to environmental hazard assessment for terrestrial vertebrates under field conditions was to apply the pesticide in question to the appropriate crop and to count the number of subsequent casualties ie. to record direct mortality. Whilst this approach does have a place in field trials, methods have been refined by the continued development of sophisticated techniques to measure residues and biochemical parameters that enable greater sensitivity in identifying the effects of exposure to the pesticide.

In the case of mammals, considerable data are already available from the medical toxicological laboratory tests for standard laboratory species. Here a sound knowledge of toxicology and their sensitivity to toxic insult permits the identification of biochemical and histological lesions which can be used to monitor field exposure (Bunyan and Stanley 1982). Where pesticides are applied to field crops, resident populations of small rodents which live on the fields provide opportunity to monitor exposure in sedentary mammals. Mice and voles are easily trapped and maintained in the laboratory, where detailed feeding studies provide the necessary data to interpret the results obtained from field specimens. Recent work at the Tolworth Laboratory of MAFF has demonstrated the value of this approach. During the course of comparative field trials with autumn sown wheat crops, populations of wood mice (Apodemus sylvaticus) have been sampled for biochemical and residue measurements following the drilling of pesticide-treated seed. An early trial with the organochlorine HCH and an organomercurial fungicide applied as seed treatments indicated a short-term pulse of detectable residues in wood mice on the field up to 14 days after drilling the grain. However residues were transitory and declined rapidly (MAFF 1978). Comparison with laboratory feeding trials showed clearly that maximum residue levels in the mice sampled on the field were considerably lower than that which produced toxic effects in laboratory feeding tests. With the replacement of the organochlorine seed treatments by organophosphorus pesticides, further intensive trials were conducted by MAFF to assess the new hazards to small rodents and birds on cereal fields. Surface grain residues were monitored after drilling, while wood mice were trapped on the study fields and in adjacent habitat and residue levels measured in their tissues. A range of enzyme activities, including plasma and tissue esterases, plasma glutamate oxaloacetate transaminase, glutamate pyruvate transaminase, glutamate dehydrogenase and sorbitol dehydrogenase, were measured in the mice (Westlake et al 1980).

Mice were found to have been exposed to residues of chlorfenvinphos for 30 days after drilling; plasma acetylcholinesterase and cholinesterase activity decreased during the first 10 days in direct relation to tissue residue levels detected in the mice. Both residues and enzyme effects were transitory however and laboratory feeding studies confirmed that this level of exposure was not a serious toxic hazard to the mice (Westlake et al 1982b). In contrast to chlorfenvinphos where residues on surface grain fell from 600 ppm after treatment to 10 ppm three days after drilling, another organophosphorus seed treatment, carbophenothion was found to be considerably more persistent. Approximately 200 ppm remained at 30 days and measurable residues were still present 6 months after drilling (Bunyan and Stanley 1979). Despite the long exposure in the field, biochemical effects were only transitory (Westlake et al 1982a) and supporting laboratory studies demonstrated that this did not present a risk of mortality (Westlake et al 1982b).

The assessment of environmental hazard from pesticides to birds has again historically been dependent on casualty searches. However the realisation of the need to measure local populations before and after treatment has led to the modified use of established census techniques. A study was conducted in which a pesticide kill was simulated by trapping and removing resident birds from a suitable field site in the spring. This clearly showed the value of census studies of breeding, territory-holding passerine birds and demonstrated the sensitivity of the method in detecting a sudden population drop as might arise from a toxic pesticide (Edwards et al 1979). A follow up study on another site confirmed the accuracy of visual and auditory censuses of breeding birds by independently comparing the results with those obtained from observations of individually marked birds. At times of year when birds are not territorial, a risk index related to numbers seen at the field trial site can be used to identify those species most at risk (Bunyan et al 1981).

Biological observations were combined with biochemical and residue measurements of sampled animals in an intensive field trial to assess the environmental effects of the carbamate, aldicarb, when it was still under development as a granular nematicide for use on sugar beet and potatoes in the UK (Bunyan et al 1981). Well established techniques for recording changes in animal populations were combined with the programmed sampling of animals, plants and soil. This demonstrated the pesticide's mobility in wet soil, that it is systemic in plants and that it can be detected widely in local fauna up to 90 days after application. This identified problems arising from granules left on the soil and the appearance of dead worms on the surface. It was concluded that between 7-30% of the birds sampled had been sufficiently exposed to aldicarb to exhibit significant signs of poisoning. In a subsequent multi-site surveillance exercise of commercial use, these hazards identified in the intensive field trial were confirmed and the application method modified to successfully overcome these problems (Bunyan et al 1981).

Measurements of residues and biochemical parameters indicated the unacceptable risk to birds presented by the experimental use of another carbamate, bendiocarb, as a seed treatment for winter wheat (MAFF 1981). Interpreting the results of the measurement of sensitive biochemical parameters in field specimens depends on the establishment of sound control data from non-exposed individuals (Westlake et al 1983) and an understanding of the factors that can affect them (Hill and Fleming 1982, Martin et al 1981, Rattner 1982).

Another intensive trial was carried out by the Tolworth Laboratory to assess the likely hazard to wildlife, particularly birds, from repeated applications of the carbamate insecticide methiocarb to ripening cherries. Nest boxes were provided throughout the orchard for resident birds to breed in and adult birds trapped and individually ringed in the orchard prior to the study. Methiocarb was applied to the orchard five times so that each tree received three sprays. Regular casualty searches were made, breeding success was monitored and birds were trapped throughout the study to provide blood samples for biochemical measurements and then released. Some birds were collected for residue monitoring. Time-related records for individual birds caught and blood sampled several times through the study supplemented the results from the programmed collection of statistically adequate samples. These showed the transience of depressed plasma acetylcholinesterase activity in individual birds which bred and moulted their body feathers successfully during the study. Although resulting in temporary biochemical changes, exposure to the pesticide did not appear to affect survival or breeding performance of birds in the orchard.

Where a pesticide has demonstrated toxicity to honeybees in the laboratory and if it is to be used on flowering crops or in other situations where bees are at risk, the hazard must be evaluated in the field. Techniques are well established in which hives are sited adjacent to fields to be sprayed and subsequent foraging behaviour, mortality and effects on the colonies are compared with observations at a control field and one treated with a toxic pesticide as a standard (ICBB 1985, Shires et al 1984a). Field methods have been further refined to examine the potential problem of bees attracted to aphid secretions in cereal crops (Shires et al 1984b). The synthetic pyrethroid insecticides demonstrate the difficulty of extrapolating from laboratory results to field hazard. High acute toxicity to bees demonstrated in the laboratory is not found in the field (Smart and Stevenson 1982) and this is thought to result from the very low field application rates and the apparent repellency of bees by fresh insecticide residues which reduces their exposure.

Earthworms are important to maintaining soil structure and fertility and are eaten by many vertebrate wildlife species. While laboratory box tests provide useful data on potential effects, there is no substitute for field evaluation where the results are more easily interpreted and more relevant to agricultural use (PSPS 1981). Earthworm mortality is measured at intervals following pesticide application and samples analysed to indicate residue levels.

Where toxicity to aquatic organisms is demonstrated in the laboratory and further field evaluation is therefore necessary, a range of test methods have been developed for pond systems in simulated field studies or for measuring the exposure from agricultural use (Hill 1985). Problems of lack of replication and adequate control are often overcome by the use of purpose-built experimental ponds or the artificial division of natural ponds. Initial populations of organisms are established and the pesticide applied before close biological monitoring is combined with residue analysis to determine the fate of the compound. Studies of natural water bodies are more difficult to conduct but techniques include, for example, confining fish in cages suspended in the water and extensive monitoring of natural populations of aquatic organisms.

POST-REGISTRATION SURVEILLANCE

A pesticide during development cannot be evaluated under all environmental conditions or against all wildlife species likely to be exposed. It is therefore important that the commercial use of a newly-introduced pesticide is closely monitored to confirm that the predictions of enviromental safety, based on limited testing, are valid in wider agricultural use. In the UK, reports of wildlife mortality thought to involve agricultural chemicals have been investigated by MAFF during the last 20 years to establish whether pesticides are responsible and to provide information to the registration authority in the event of an unacceptable hazard requiring change of clearance status (Hardy and Stanley 1984, Hardy et al 1986). Two good examples illustrate the value of this approach. When first introduced into the UK, carbophenothion used as a cereal seed treatment presented an unacceptable risk to wintering wildfowl (Stanley and Bunyan 1979) and the granular soil insecticide aldicarb presented a hazard to gulls. Both problems were successfully overcome by reducing the exposure of wildlife species at risk to the particular pesticide. Carbophenothion was withdrawn from use in those parts of the UK where wildfowl, particularly geese, winter in greatest numbers. In the case of aldicarb, application equipment was modified and advice to farmers strengthened on environmental risks from poor soil incorporation of granules. Through joint agreement of industry and government, these two useful agricultural pesticides have been retained on the UK market with a significant reduction in environmental hazards to wildlife.

The evaluation of the environmental risks presented by a new pesticide therefore involves three stages. On the basis of the laboratory identification of the intrinsic properties of a compound, a sound field testing programme is developed and carried out tailored to the intended use of the pesticide. Post-registration surveillance of commercial use provides the assurance that predictions of environmental safety, from the results of testing, hold true in agricultural use. This approach in the UK has confirmed few significant problems arising from the agricultural use of new pesticides. This has added confidence to the process of rigorous environmental hazard assessment of new pesticides by laboratory and field tests.

REFERENCES

Balcomb R, Stevens R, Bowen C (1984) Toxicity of 16 granular insecticides to wild-caught songbirds. Bull Environm Contam Toxicol 33: 302-307.

Bunyan PJ, Stanley PI (1979) Assessment of the environmental impact of new pesticides for regulation purposes. Proc Br Crop Prot Conf Pests Dis, 10th, 881-891.

Bunyan PJ, Stanley PI (1982) Toxic mechanisms in wildlife. Regul Toxicol Pharmacol 2: 106-145.

Bunyan PJ, Van den Heuvel MJ, Stanley PI, Wright EN (1981) An intensive field trial and a multi-site surveillance exercise on the use of aldicarb to investigate methods for the assessment of possible environmental hazards presented by new pesticides. Agro-Ecosystems 7: 239-262.

Edwards PJ, Brown SM, Fletcher MR, Stanley PI (1979) The use of a bird territory mapping method for detecting mortality following pesticide application. Agro-Ecosystems 5: 271-282.

EPA (1985) Soil Column Leaching. Hazard Evaluation Division Standard Evaluation Procedure. Office of Pesticide Programmes, Environmental Protection Agency, Washington.

FAO (1981) Second expert consultation on environmental criteria for registration of pesticides. FAO PLant Production and Protection Paper 28. Food and Agriculture Organisation of the United Nations, Rome.

FAO (1982) Report of the Second Government Consultation on International Harmonization of Pesticide Registration Requirements. Food and Agriculture Organisation of the United Nations, Rome (AGP 1982/M/5).

Fink RJ, Jaber MJ (1981) The laughing gull (Larus atricilla) as a model for the assessment of secondary poisoning. In Avian and Mammalian Wildlife Toxicology: 2nd Conference ASTM STP 757 (eds) Lamb DW, Kenaga EE American Society Testing and Materials: 66-71.

Hardy AR, Stanley PI (1984) The impact of the commercial agricultural use of organophosphorus and carbamate pesticides on British wildlife. In: Agriculture and the Environment, (ed) Jenkins D, 72-80. (ITE Symposium No.13). Cambridge, Institute of Terrestrial Ecology.

Hardy AR, Fletcher MR, Stanley PI (1986) Twenty years of vertebrate wildlife incident investigations by MAFF. Veterinary Journal (in press).

Hassan SA, Bigler F, Blaisinger P, Bogenschutz H, Brun J, Chiverton P, Dickler E, Easterbrook MA, Edwards PJ, Englert WD, Firth SI, Huang P, Inglesfield C, Klingauf F, Kuhner C, Ledieu MS, Naton E, Oomen PA, Overmeer WPJ, Plevoets P, Reboulet JN, Rieckmann W, Samsoe-Petersen I, Shires SW, Staubli A, Stevenson JH, Tuset JJ, Vanwetswinkel G, Van Zon AQ (1985) Standard methods to test the side-effects of pesticides on natural enemies of insects and mites developed by the IOBC/WPRS Working Group 'Pesticides and Beneficial Organisms'. EPPO Bulletin 15: 214-255.

Hill EF, Fleming WJ (1982) Anticholinesterase poisoning in birds: field monitoring and diagnosis of acute poisoning. Environ Toxicol Chem 1: 27-38.

Hill EF, Heath RG, Spann JW, Williams JD (1975) Lethal dietary toxicities of environmental pollutants to birds. US Fish Wildl Serv, Spec Sci Rep - Wildl 191, 61pp.

Hill EF, Carmadese MB (1986) Lethal dietary toxicity of environmental contaminants and pesticides to coturnix. US Fish Wildl Serv Tech Rep 2, 147pp.

Hill IR (1985) Effects on non-target organisms in terrestrial and aquatic environments. In: The Pyrethroid Insecticides (ed) Leahey JP, 151-262. Taylor and Francis, London Philadelphia.

Hoffman DJ, Albers PH (1984) Evaluation of potential embryotoxicity and teratogenicity of 42 herbicides, insecticides and petroleum contaminants to mallard eggs. Arch Environ Contam Toxicol 13: 15-27.

Hornshaw TC, Aulreich RJ, Ringer RK (1986) Toxicity of orthocresol to the European ferret. Environ Toxicol Chem 5: 713-20.

Hudson RH, Tucker RK, Haegele MA (1984) Handbook of toxicity of pesticides to wildlife (2nd ed). US Fish Wildl Serv Res Pub 153, 90pp.

International Commission for Bee Botany (1985) Third symposium on the harmonisation of methods for testing the toxicity of pesticides to bees, 18-21 March 1985, Rothamsted Experimental Station England (ed) Felton JC 11pp, 26 appendices.

MAFF (1978) Pest Infestation Control Laboratory Report 1974-76. London: HMSO.

MAFF (1981) Agricultural Science Service Pesticide Science 1979 (Reference Book 252) London HMSO.

MAFF (1986) Data requirements for approval under the Control of Pesticides Regulations 1986. Ministry of Agriculture, Fisheries and Food, Pesticides Branch, London.

Martin AD, Norman G, Stanley PI, Westlake GE (1981) Use of reactivation techniques for the differential diagnosis of organophosphorus and carbamate pesticide poisoning in birds. Bull Environm Contam Toxicol 26: 775-780.

OECD (1981) Chemical Group OECD Test Guidelines. Expert Group on Physical Chemistry, May 1981.

PSPS (1981 Laboratory and field testing of pesticide products for effects on soil macro-organisms. Pesticides Safety Precautions Scheme Working Document D6.

Rattner BA (1982) Diagnosis of anticholinesterase poisoning in bird: effects of environmental temperature and underfeeding on cholinesterase activity. Environ Toxicol Chem 1: 329-336.

Schafer EW (1972) The acute oral toxicity of 369 pesticidal, pharmaceutical and other chemicals to wild birds. Toxicology and Applied Pharmacology 21: 315-330.

Schafer EW, Bowles WA, Hurlbut J (1983) The acute oral toxicity, repellency and hazard potential of 998 chemicals to one or more species of wild and domestic birds. Arch Environ Contam Toxicol 12: 355-382.

Shires SW, Le Blank J, Murray A, Forbes S, Debray P (1984a) A field trial to assess the effects of a new pyrethroid insecticide, WL85871, on foraging honeybees in oilseed rape. J Apicult Res 23: 217-226.

Shires SW, Le Blanc J, Debray P, Forbes S, Louveaux J (1984b) Field experiments on the effects of a new pyrethroid insecticide WL85871 on bees foraging artificial aphid honeydew on winter wheat. Pestic Sci 15: 543-552.

Smart LE, Stevenson JH (1982) Laboratory estimation of toxicity of pyrethroid insecticides to honeybees: relevance to hazard in the field. Bee Wld 63: 150-152.

Somerville L, Greaves MP, Domsch KH, Verstraete W, Poole NJ, Van Dyk H, Anderson JPE (1986) Recommended laboratory tests for assessing the side effects of pesticides in the soil microflora (Proceedings of the 3rd International Workshop, Cambridge). Institute of Arable Crops Research, Long Ashton Research Station, Bristol.

Stanley PI, Bunyan PJ (1979) Hazards to wintering geese and other wildlife from the use of dieldrin, chlorfenvinphos and carbophenothion as wheat seed treatments. Proc R Soc Lond B 205: 31-45.

Stanley PI, Hardy AR (1983) Methods of prediction of environmental effects of pesticides. Field trials to assess the hazard presented by pesticides to terrestrial wildlife. In: Proc Int Congr Plant Protection, 10th, Brighton, 1983, 2: 692-701. Croydon: British Crop Protection Council.

Stevenson JH (1968) Laboratory studies on the acute contact and oral toxicity of insecticides to honeybees. Ann Appl Biol 61: 467-472.

Townsend MG, Fletcher MR, Odam EM, Stanley PI (1981) An assessment of the secondary poisoning hazard of warfarin to Tawny Owls. J Wild Manage 45: 242-247.

Townsend MG, Odam EM, Stanley PI, Wardall HP (1983) Assessment of secondary poisoning hazard of warfaring to weasels. J Wild Manage 48: 628-632.

Tucker RK, Haegele MA (1971) Comparative acute oral toxicity of pesticides to six species of birds. Toxicol Appl Pharmacol 20: 57-65.

Tucker RK, Leitzke JS (1979) Comparative toxicology of insecticides for vertebrate wildlife and fish. Pharmacol Therap 6: 167-220.

Westlake GE, Blunden CA, Brown PM, Bunyan PJ, Martin AD, Sayers PE, Stanley PI, Tarrant KA (1980) Residues and effects in mice after drilling wheat treated with chlorfenvinphos and an organomercurial fungicide. Ecotoxicol Environ Saf 4: 1-16.

Westlake GE, Brown PM, Bunyan PJ, Felton CL, Fletcher WJ, Stanley PI (1982a) Residues in mice after drilling wheat treated with carbophenothion and an organomercurial fungicide. In: Environmenta and Quality of Life 522-527 (Proc Int Symp Principles for the Interpretation of the Results of Testing Procedures in Ecotoxicology 1980) Luxembourg: Commission of the European Communities.

Westlake GE, Bunyan PJ, Johnston JA, Martin AD, Stanley PI (1982b) Biochemical effects in mice following exposure to wheat treated with chlorfenvinphos and carbophenothion under laboratory and field conditions. Pestic Biochem Physiol 18: 49-56.

Westlake GE, Martin AD, Stanley PI, Walker CH (1983) Control enzyme levels in the plasma, brain and liver from wild birds and mammals in Britain. Comparative Biochemistry and Physiology 76: 15-24.

CLINICAL MANAGEMENT OF PESTICIDE POISONING

L. Rosenstock MD, MPH
Departments of Medicine and Environmental Health
University of Washington, ZA-66
Seattle, Washington 98195

Pesticide intoxication, whether by intentional ingestion or accidental overexposure, remains a significant problem in both industrialized and developing countries [1]. Although the nature and extent of exposure to the offending agent commonly may be known when caring for a patient with pesticide poisoning, there continue to be serious problems with the awareness, recognition, and appropriate diagnosis and treatment of these intoxications [2]. Presented below is an approach to the clinical management of patients with suspected or known pesticide overexposure, grouped into the major chemical classes of pesticides. In evaluating the potential health-related effects of all classes of pesticides, it is important to think not only of the active ingredients but also of the carriers, which are often added to enhance delivery; carriers include talcs, oils, solvents and binding agents. These so-called inert ingredients may themselves have toxic properties explaining some of the acute and persistent health effects, acting as irritants and less commonly sensitizing agents in exposed individuals.

INSECTICIDES

Organophosphates

Because of the attractiveness of these agents in terms of their short-lived residues and minimal persistence in the environment, these chemicals have become of increasing commercial importance. Concomitant with their more prominent role as pesticides is an increase in intoxications with these agents [1,3,4]. As a class, organophosphates have generally high acute toxicity. The main clinical effects in humans are mediated through the inhibition of the enzymes acetylchlinestrase in both the central and peripheral nervous system. The expression of these effects is due to the subsequent buildup of acetylcholine at neural synapses. Dermal, respiratory, and gastrointestinal tract absorption are all potential routes of intoxication, although absorption via the skin is the most significant and usual route of occupational intoxication. Most effects of clinical concern result from systemic toxicity, but because some of these compounds are direct cholinesterase inhibitors (not requiring metabolic oxidation for activity) local effects at the site of contact may also be evident [4]. Such local effects include miosis and blurred vision as well as skin irritation, muscle fasciculation, and focal diaphoresis. Skin and respiratory sensitization are uncommon but can develop from reexposure to some of these agents.

CLINICAL EFFECTS
Acute

The symptoms and signs of acute systematic intoxications, arising from excessive acetylcholine at central and peripheral sites, can be classified into three categories of nervous system involvement [2,4,5].

1. Muscarinic Effects
These parasympathetic effects result from cholinergic activity at smooth muscle and exocrine gland sites and are usually the first signs and symptoms of poisoning. Muscarinic signs and symptoms include the following: anorexia, nausea and vomiting, abdominal cramps, diarrhea, chest tightness (bronchoconstriction and increased secretions), increased salivation and lacrimation, miosis and blurred vision, diaphoresis, bradycardia, and involuntary defecation and urination in cases of severe intoxication.

2. Nicotinic Effects
Sympathetic and somatic motor effects are mediated by increased cholinergic activity of nerves innervating skeletal muscle and autonomic ganglia. Nicotinic effects include the following: muscle twitching, fasciculation, weakness, hypertension, and hyperglycemia. Tachycardia may occur in cases of severe intoxication, overriding the muscarinic-mediated bradycardia.

3. Central Nervous System Effects
These neurologic effects are mediated through cholinergic activity in the central nervous system. Central nervous system effects include

anxiety, irritability, altered sleep patterns, headache, tremor, impaired cognition, and seizures and coma in the cases of severe intoxication.

In general, the onset of symptoms and signs from organophophate intoxication occurs more rapidly following inhalational rather than gastrointestinal or skin absorption. Onset of effects may range from minutes to hours, but there is significant variation in the onset of symptoms depending on the nature of the chemical and its metabolism as well as variations in individual responses. Delayed onset of symptoms and signs has been reported as far as 48 hours after exposure.

Delayed Neurological Effects Of Acute Intoxication

Independent of effects of cholinesterase inhibition is the recognized axonopathy that occurs from a few chemicals in this class [6]. These effects have been attributed to inhibition of neurotoxic esterases. The prototype of this process occurs with overexposure to triorthocresyl phosphate (TOCP). Clinical presentation may occur days to weeks following the acute exposure and is characterized by ataxia and a mixed motor-sensory disturbance, first affecting the lower extremities. When severe, motor complications may result in a flaccid paralysis.

Chronic

The chronic sequelae of acute intoxication are much less well defined than the acute clinical manifestations of these exposures. Although not the subject of this discussion, the clinician managing patients with acute intoxication should be aware that chronic neuropsychologic effects may result [7,8]; appropriate clinical follow-up, including neuropsychologic testing, should be undertaken in individuals who do not show full recovery within days to weeks following an episode of acute intoxication.

EVALUATION AND TREATMENT

As with other causes of pesticide intoxication, the history of exposure is germain to the appropriate diagnosis and treatment of organophosphate toxicity. Wherever possible, a description of the specific chemical should be obtained. Product label information is helpful, as well as a description of the setting of exposure and probable absorption route(s). As most of the acute toxic effects are mediated through the inhibition of cholinesterase activity with resultant cholinergic effects, physical examination findings should be directed to the constellation of signs outlined earlier. Of most utility is the examination of vital signs, with bradycardia the most common finding, although in more severe intoxication tachycardia may predominate. Miosis of the pupils, diaphoresis, and signs and symptoms of pulmonary congestion and bronchospasm are also important clues. Laboratory evaluation is directed at indirect measures of central and peripheral nervous system cholinesterase inhibition, through measurement of red blood cell (RBC) cholinesterase (acetylcholinesterase) and plasma cholinesterase (pseudocholinesterase) [9]. In cases of suspected acute toxicity, both of these levels should be obtained; if these are not available, measurement of whole blood cholinesterase provides a reasonable alternative. The RBC cholinesterase level parallels more closely nervous system cholinesterase inhibition than does the plasma level, but it may be depressed less rapidly than plasma cholinesterase so that normal RBC and low plasma levels are consistent with early phases of acute toxicity. This pattern of normal RBC and low plasma cholinesterase levels may also be seen with early acute carbamate intoxication (see later section) as well as the two following situations [2,9]: 1) genetic pseudocholinesterase deficiency and 2) acquired cholinesterase deficiency, which may occur in altered production states such as chronic hepatic disease. In organophosphate intoxication, plasma cholinesterase levels are likely to return to normal within two weeks following exposure, but red blood cell depression may persist over several months. Patterns of cholinesterase levels in organophosphate and carbamate intoxication are outlined in Table 1.

TABLE 1   PATTERNS OF CHOLINESTERASE ACTIVITY

Setting	Acetylcholinesterase (RBC) Levels	Pseudocholinesterase (Plasma) levels
Within hours of of acute organo- phosphate carbamate intoxication	Decreased or Normal	Decreased
Acute organo- phosphate Intoxication (late)*	Decreased	Decreased
Acute carbamate Intoxication (late)*	Normal	Normal
Genetic Pseudo- cholinesterase deficiency	Normal	Decreased
Chronic liver disease and pregnancy	Normal	Decreased

*   at least several hours following exposure

The treatment of acute organophosphate intoxication includes two specific measures aimed at reversing the effects of cholinesterase inhibition.   The first of these is atropinization, which exerts anti-cholinergic effects on parasympathetic receptors.   In cases of severe intoxication, patients can be given 2-4 mg of atropine sulfate intravenously with repeated doses every 5 to 10 minutes until signs of atropinization are achieved (drying of mucous membranes, skin, and bronchial secretions and induction of tachycardia).   The necessary dose to achieve atropinization is highly variable and in extreme cases may necessitate several grams administered over several days. The second specific measure, employed in conjunction with atropinization, is administration of 2-PAM (pralidoxime), initially given intraveneously as a 1 gm dose in 250 ml of saline.   This dose may be repeated if significant clinical improvement is not observed and, in cases of severe intoxication, can then be administered at an infusion rate of .5 grams/hour.   Pralidoxime acts as a specific antidote by disrupting the acetyl-cholinesterase-organoposphate bond.

In addition to these specific measures, supportive therapy is key and should include protection and support of airway and ventiliation as indicated.   Finally, decontamination by removal of clothing and soaping of skin should be undertaken as soon as possible. This measure should be undertaken with caution to avoid intoxication of others from contact with the pesticide on a patient's skin and clothing.   Individuals occupationally exposed to pesticides who suffer acute intoxication should not return to activities with potential reexposure until cholinesterase levels have returned to normal. In addition, these individuals should be monitored for recurrence of symptoms and signs of repeated overexposure upon return to work.

CARBAMATES

Because these substances tend to have generally lower level of toxicity than organophosphates, they have become increasingly used as insecticides, as well as fungicides and herbicides.   Nonetheless, the range of toxicity of these substances is still broad such that some chemicals in this class are as toxic both orally and dermally as the most toxic organophosphates. Carbamates do not form irreversible bonds with the cholinesterase enzyme, but acute toxicity is mediated through the irreversible bonds formed with this enzyme and resultant cholinesterase inhibition and cholinergic effects.

CLINICAL EFFECTS

Acute effects are indistinquishable from those caused by organophosphates. The onset of action of these effects, however, is generally more rapid than with organophosphate intoxication. Hence, individuals facing overexposure to carbamates may remove themselves from further exposure and sustain lower overall doses than those with organophosphate intoxication. Delayed neurotoxic effects have not been reported with this class of insecticides and chronic sequelae are even less well studied than those with organophosphates.

EVALUATION AND TREATMENT

Clinical evaluation is the same as that for organophosphate intoxication. Laboratory evaluation is more problematic as both plasma and RBC cholinesterase levels may return to normal within hours of acute intoxication. In instances where an individual is seen hours from acute exposure, diagnosis and treatment will often rely on identification of exposure, clinical signs, and response to therapy. In addition, as with organophosphate intoxication, the documentation of exposure can be achieved by attempts to identify carbamate compounds in blood or their metabolites in urine [10]. The availability of these tests is limited, however, and they are not routinely needed in clinical management.

In addition to supportive therapy, which includes decontaimination and maintenance of airway and ventilation, the specific treatment of atropinization is indicated in the setting of carbamate intoxication. Unlike treatment for organophosphates, however, pralidoxime is not indicated; the action of this agent is specific to the phosphorylated enzyme, not present in the carbamate-intoxicated individual. Moreover, some individuals would consider pralidoxime contraindicated because in and of itself it may exert weak cholinesterase inhibiting effects.

ORGANOCHLORINE INSECTICIDES

These insecticides fall into three main groups: the chlorinated ethane compounds (such as DDT), the cyclodienes (such as chlordane), and the hexachloro-cyclohexanes (such as lindane). The mechanism of the neurotoxicity effects of these compounds is not well understood and in general these compounds are less acutely toxic than the cholinestrase inhibitors. Routes of concern include inhalation and ingestion for all chemicals in this class, with rates of dermal absorption varying from minimal to high.

CLINICAL EFFECTS

Acute

Most of the effects observed in the setting of acute intoxication are central nervous system ones. These effects include headache, nausea, dizziness and irritability. In more severe intoxication the following may be observed: weakness, ataxia, tremor, confusion, fasciculations, and central and peripheral paresthesias. In addition, seizures and coma may be seen in intoxication with these agents, and may be the presenting sign of intoxication with some substances of this class, such as cyclodiene poisoning [11].

EVALUATION AND TREATMENT

The approach to clinical evaluation of the exposure history and physical examination is similar to that described above for intoxication with cholinestrase inhibitors. Laboratory testing of exposure, however, is more limited; there are no available biological markers, such as cholinestrase levels, to monitor the nature and extent of intoxication. Specialized laboratory procedures are available for measuring the parent compounds or metabolites in body tissues, and should be reserved for those instances when the source of intoxication is not otherwise evident.

The treatment of intoxication by this class of insecticides is confined to decontamination and supportive measures. There are no specific antidotes. In addition, as with seizures from other causes, antiepileptiform medications should be initiated if seizures are observed. Some of these substances are known to be hepatotoxic, but again no specific therapy for this effect is available. An individual demonstrating hepatoxicity should be monitored for resolution of functional abnormalities, which may persist for months following acute intoxication.

HERBICIDES

The three main classes of compounds in this class include the chlorophenoxy compounds, the dinitrophenols, and the bipyridyl compounds. Toxicity from exposure to these compounds is highly varied. In the case of chlorophenoxy compound exposure, many of the effects have been attributed to their frequent contamination with dioxins, although there is evidence of peripheral neuropathic effects from exposure to these compounds independent of such contamination [12]. Unique among these compounds is paraquat, a bipyridyl herbicide with acute systemic absorption associated with pulmonary injury that may be severe and fatal as well as acute renal failure [13]. Another agent of this class, diquat, has not been associated with pneumonitis, but has been associated with hepatic and renal failure.

Treatment of intoxication with all these agents is supportive, with no specific antidotes available.

FUNGICIDES

The class of fungicides constitutes a heterogeneous group of chemicals which in general have minimal acute toxicity. Exceptions include the organomercury compounds which are potent neurotoxins, hexachlorobenzene which is a strong porphyrinogen, and pentachlorophenol (PCP) which like the dinitrophenol-herbicides can increase the basal metabolic rate through decoupling of cellular respiration (1). Resultant symptoms and signs of intoxication include tachycardia, hyperthermia, diaphoresis, and tachypnea. Concomitant hepatic and renal toxicity have also been observed in cases of acute intoxication. Skin, mucous membrane, and respiratory tract irritation may also arise from exposure to PCP.

Treatment of acute intoxication with these agents is generally supportive, including decontamination and other measures. A forced diuresis has been employed to enhance PCP clearance in cases of acute intoxication with this substance.

SUMMARY AND CONCLUSION

This discussion has focused mainly on acute intoxication with the cholinesterase inhibitors, the organophosphates and carbamates, not only because acute toxicity may be significant but because specific evaluation and therapeutic interventions are available and useful.

For all pesticide intoxications, however, a general approach for the clinician is to maintain awareness about the possibility of these agents as etiologic in occupational and other environmental settings. For all, identification of overexposure relys heavily on the history and examination, including identification of the agent when this information is available. Laboratory testing can be applied in specific settings. Treatment of all causes of pesticide intoxications includes decontamination and supportive therapy. Monitoring for resolution of acute intoxication and return to normal of altered laboratory function should guide subsequent follow-up care, and is particularly important in the occupational setting as individuals return to potential reexposure to these agents.

REFERENCES

1.   Wasserman RF, Wiles R (1985) Field duty: U.S. farmworkers and pesticide safety. World Resources Institute Study 3
2.   Rostenstock L and Cullen MR (1986)   Clinical Occupational Medicine WB Saunders Co Philadelphia
3.   Midtling JE, Barnett PG, Coye MJ et al (1985) Clinical management of field worker organophosphate poisoning. West J Med 142:514-18
4.   Murphy SD. Toxic effects of pesticides. In Doull J, Klaassen CD, Amdur MO (eds) (1980). Toxicology, 2nd edition. Macmillan Publishing Co. New York
5.   Hamba T, Nolte CT, Jocknel J et al (1971). Poisoning due to organophosphate insecticides: acute and chronic manifestations. Am J Med 50:475-92
6.   Senanayake H, Johnson MK (1982) Acute polyneuropathy after poisoning by a new organophosphate insecticide. N Engl J Med 306: 155-57
7.   Durham WF, Wolfe HR, Quinby GE (1965) Organophosphate insecticides and mental alertness. Arch Environ Health 10:55-56
8.   Duffy FH, Burchfield JL, Bartels PH et al (1979) Long-term effects of an organophosphate upon the human electroencephalogram. Tox and Appl Pharm 47:161-76.
9.   Coye MJ, Lowe JA, Maddy KT (1986) Biological monitoring of agriculatural workers exposed to pesticides: I. Cholinesterase activity determinations. J Occup Med 28: 619-27
10. Coye MJ, Lowe JA, Maddy KJ (1986) Biological monitoring of agricultural workers exposed to pestidies: II. Monitoring of intact pesticides and their metabolites. J Occup Med 28:628-36
11. Davies GM, Lewis I (1956) Outbreak of food poisoning from bread made from chemically contaminated flour. Br Med J 2:393-98
12. Berkley MC, Magee KR (1963). Neuropathy following exposure to a dimethylamine salt of 2,4-D. Arch Intern Med 111:351-52
13. Davies DS, Hawksworth GM, Bennett PN (1977) Paraquat poisoning. Proc Eur Soc Toxicol 18:21-26

ASSESSMENT OF HUMAN EXPOSURE TO PESTICIDES

Marcello Lotti

Istituto di Medicina del Lavoro

Universita' degli Studi di Padova

Via Facciolati 71, 35127 Padova, Italy.

The aim of the assessment of human exposure to pesticides is the definition of dose-effect relationships in man after both single and repeated exposures and therefore the prevention of the adverse effects due to these chemicals.

According to the circumstance, size of dose, and methods of assessment, human exposures might be divided as: a. acute/subacute poisonings (intentional, accidental, occupational), b. long term occupational exposures and c. environmental exposures (via food, water etc). The relationship among different exposures is depicted in Figure 1.

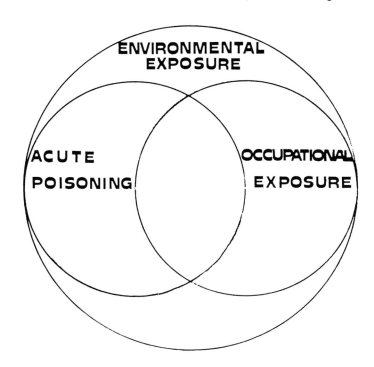

NATO ASI Series, Vol. H13
Toxicology of Pesticides: Experimental, Clinical and Regulatory Aspects. Edited by L. G. Costa et al.
© Springer-Verlag Berlin Heidelberg 1987

It should be noted however that no comprehensive information exists on the exposure of the world population to pesticides, neither in the occupational nor in the general environment (1).

Toxicological, clinical and epidemiological criteria are used in the assessment of human exposures to pesticides.

Toxicological aspects in the evaluation of human exposures to pesticides

Toxicological studies are the cornerstone for an assessment of human exposures to pesticides. They can be divided into two main groups: those which address the problems of the delivery of toxic chemicals to the site of action and those which study the interaction of the toxic chemical with the molecular target and the reactions which follow.

Figure 2 shows the fate and reactions of a xenobiotic in the human body and the types of tests available to investigate human exposures (2,3).

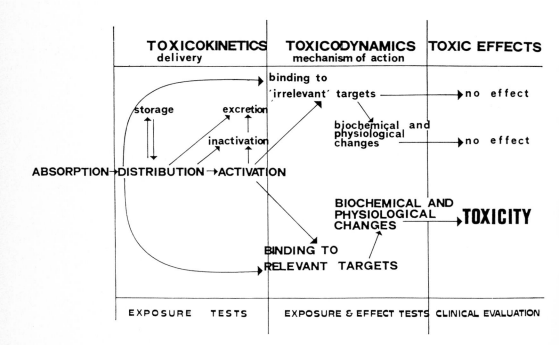

The lower part of the figure describes the chain of events leading to toxic effects. Once a chemical is absorbed and distributed through the plasma pool, it attaches itself to the molecular target either directly or after metabolic activation. Following this interaction, a cascade of biochemical and physiological changes occurs, which triggers the morphological/ clinical expression of toxicity. Evaluations of human exposures might be performed at any stage of this process but their significance is obviously different.

Tests exploring the toxicokinetics include measurements of the chemical or of its metabolites in body fluids and their aim is the definition of the dose. Virtually all pesticide exposures can be assessd in this way, only depending on availability of analytical procedures. Several examples are given in the comprehensive textbook by Hayes (4). Exposure tests have also been recently derived from the pathway which involves the binding to "irrelevant" targets, as shown in the upper part of the figure. For instance, haemoglobin adducts, as measured in red blood cells, can be used as a more precise determination of the internal dose (5,6). Furthermore, the meausurement of DNA adducts after exposures to various chemicals and pesticides should be regarded as an exposure test to potential carcinogens (7) even though the relationship between mutagenicity and chemical carcinogens is still unclear (8), and the significance of DNA adducts not fully understood (9).

The binding to the relevant targets and their consequences represent the mechanism of action of a chemical and when it is understood, as in the case of several pesticides, specific effect tests can be derived. A recent example includes the understanding of the mechanism of action of organophosphates causing delayed polyneuropathy (10,11) and the development of a specific clinical test to evaluate this toxicity (12,13). Furthermore, the understanding of the mechanism of action and the availability of an effect test allows studies on quantitative relationships between the concentration of a chemical or its metabolites in body fluids and the effect on the targets. This is known, for instance, for the pesticide parathion by measuring p-nitrophenol urinary excretion and inhibition of RBC acethylcholinesterases (14).

It should be kept in mind however that toxicokinetics and toxicodynamics can be altered in particular circumstances of exposure and therefore special care must be exercised in the interpretation of exposure and effect tests. For instance, as it often occurs in pratical conditions of exposure to impurities and mixtures of pesticides, the toxicokinetics might be deeply influenced and the "net" toxicological effect remarkably changed. Examples include inhibition of the detoxification mechanisms as in the case of impurities of organophosphates (15,16) or as in the case of mixtures of pyrethroids and organophosphates (17), and also accelerated biotransformation as in the case of liver enzyme induction by certain chlorinated pesticides (18,19). Toxicodynamics might also be influenced by the competition for the target as in the case of some carbamates which prevent the delayed polyneuropathy caused by some organophosphorus esters (10,11). Furthermore, in the case of repeated exposures, tolerance to toxic effects might develop. Examples include the tolerance to the anticholinesterase effects of some organophosphates after repeated exposures, which seems to be related to the "down regulation" of acetylcholine receptors (20).

In conclusion, human exposures to pesticides might require a very complex evaluation but when exposure and effect tests are available the toxicological picture can be rationalized.

Clinical evaluation

The clinical evaluation of the toxic effects of pesticides usually follows high exposure (21). The effects on the nervous system are the prominent ones for most pesticides and include those of chlorinated hydrocarbon insecticides, pyrethroids and the well known cholinergic ones of organophosphates & carbamates. Major clinical effects on other organs are less common in acute pesticide poisoning but, for instance, severe damage to the lung is caused by bipyridylium herbicides. A comprehensive review of all reported human cases of pesticide poisoning is available (4).

The basis for reliable clinical reports is the identification of the chemical(s) and an estimate of the dose the patient(s) was exposed to. Therefore, the recovery of the original chemical (bottles, cans etc) would be extremely useful whenever possible. In fact, the presence of other chemicals and impurities might substantially contribute to the clinical picture as mentioned above. Another aspect which is often overlooked in emergency rooms is the recovery and appropriate storage of gastric lavage and dialysis fluids for future chemical analysis. Serial sampling of blood and urine might provide elements for both the identification of the chemical and for pharmacokinetic studies. Very few "effect tests" are available to be used for diagnostic purposes of pesticide poisoning; a specific diagnostic test for the acute effects of organophosphate and carbamate pesticides is the measurement of red blood cell acetylcholinesterase inhibition which reflects, in most circumstances, that in the target organ. Particular care should be used in performing and evaluating this test in the case of carbamates which are reversible inhibitors of acetylcholinesterases (4). Measurements of plasma pseudocholineterases, a test which should be distinguished from the acetylcholineterase one, might be useful as an "exposure" test, as this enzyme represents one of the "irrelevant targets" of organophosphates (4). For the rare effect of organophosphate delayed polyneuropathy, an "effect test" is at the validation stage and has been demonstrated diagnostic for this toxicity (12,13).

The current diagnostic procedures which are used in general medical practice should be evaluated in that particular poisoning case in terms of specificity and sensitivity.

The clinical evaluation of non-acutely poisoned exposed subjects might also be performed by means of electrophysiological studies of the central and peripheral nervous system because these tecniques might be more sensitive.

The clinical assessment associated to the toxicological studies (to measure the dose) provides key information on the threshold for clinical/sub-clinical effects (22).

Epidemiological evaluation

Epidemiological studies are equally necessary in investigations of outbreaks of toxic poisoning, in the monitoring of occupational exposures and in the evaluation of low environmental exposures to pesticides.

Environmental carcinogenesis is perhaps the area of greatest concern at the moment and epidemiological studies have greatly contributed to our knowledge; however, no data are available on pesticides (1). The Directory of on-going Research in Cancer Epidemiology in 1985 lists only 19 studies in the world on pesticides, and most of them related to phenoxyacid exposures (23).

Major difficulties in performing epidemiological studies on pesticide exposures include the assessment of the dose, the variability and unpredictability of agricultural exposures, and the occupational medicine practice which is less implemented in agriculture than in the industry.

Major efforts in analytical chemistry, occupational and clinical toxicology, and epidemiology might lead to a better assessment of human exposures and to the prevention of the toxic effects of pesticides.

References

1. International Agency for Research on Cancer Monographs.(1983) Miscellaneous Pesticides. Vol.30, Lyon.
2. Aldridge W.N. (1981) Mechanisms of toxicity. New concepts are required in toxicology.Trends in Pharmacological Sciences. 2, 228-231.
3. Aldridge W.N. (1986) The biological basis and measurement of thresholds. Annual Reviews in Pharmacology and Toxicology 26, 39-58.
4. Hayes W.J.Jr. (1982) Pesticides studied in man. Williams & Wilkins. Baltimore.
5. Neumann, H.G. (1984) Analysis of haemoglobin as a dose monitor of alkylating and acylating agents. Archives of Toxicology 56, 1-6.
6. Albrecht W. and Neumann H.G. (1985) Biomonitoring of aniline and nitrobenzene: haemoglobin binding in rats and analysis of adducts. Archives of Toxicology 57, 1-5.
7. Vainio H. (1985) Current trends in the biological monitoring of exposure to carcinogens. Scandinavian Journal of Work and Environmental Health 11, 1-6.
8. International Agency for Research on Cancer (1980). Long-term and short-term screening assays for carcinogens: a critical appraisal. Supplement 2. Lyon.
9. Maugh T.H. (1984) Tracking exposure to toxic substances. Science 226, 1183-1184.
10. Johnson M.K. (1982) The target for initiation of delayed neurotoxicity by organophosphorus esters: biochemical studies and toxicological applications. Reviews in Biochemical Toxicology 4, 141-212.
11. Lotti M., Becker CE., Aminoff M.J. (1984) Organophosphate polyneuropathy: pathogenesis and prevention. Neurology, 34, 658-662.
12. Lotti M., Moretto A., Zoppellari R., Dainese R., Rizzuto N., Barusco G. (1986) Inhibition of lymphocytic Neuropathy Target Esterase predicts the development of organophosphate induced delayed polyneuropathy. Archives of Toxicology, 59, 176-179.
13. Lotti M. (1987) Organophosphate-induced delayed polyneuropathy in man and perspectives for biomonitoring. Trends in Pharmacological Sciences, in press.
14. Arterberry J.D., Durham W.F., Elliott J.W., and Wolfe H.R. (1961) Exposure to parathion. Measurement by bloood cholinesterases level and urinary p-nitrophenol excretion. Archives Environmental Health. 3, 476-485.
15. Baker E.L., Zack M., Miles J.W., Alderman L., McWilson W., Dobbin R.D., Miller S., Teeters W.R. (1978) Epidemic Malathion Poisoning in Pakistan Malaria Workers. Lancet, 1, 31-33.
16. Aldridge W.N., Miles J.W., Mount D.L., Verschoyle R.D. (1979) The toxicological properties of impurities in Malathion. Archives of Toxicology, 42, 95-106.

17. Miyamoto J. (1976) Degradation, metabolism and toxicity of synthetic pyrethroids. Environmental Health Perspectives 14, 15-28.

18. Kolmodin B., Azarnoff D.L., Sjoqvist F. (1969) Effect of environmental factors on drug metabolism: decreased plasma half-life of antipyrine in workers exposed to chlorinated hydrocarbon insecticides. Clinical Pharmacology and Therapeutics 10, 638-642.

19. Poland A., Smith D., Kuntzman R., Jacobson M., Conney A.H. (1970) Effect of intensive occupational exposure to DDT on phenylbutazone and cortisol metabolism in human subjects. Clinical Pharmacology and Therapeutics 11, 724-732.

20. Costa L.G., Schwab B.W., Murphy S.D.(1982) Tolerance to anticholinesterase compounds in mammals. Toxicology, 25, 79-97.

21. Plestina R. (1984) Prevention, diagnosis and treatment of insecticide poisoning. World Health Organization, VBC/84.889.

22. Gilioli R., Cassitto M.G., Foa' V. (1983) Neurobehavioral Methods in Occupational Health. Advances in the Biosciences. Volume 46, Pergamon Press.

23. International Agency for Research on Cancer (1985) Directory of on-going research in cancer epidemiology 1985. Lyon.

PESTICIDES - RISK ASSESSMENT AND SAFE USE

T B Hart, Products Medical Adviser, ICI Plant Protection Division

INTRODUCTION

The use of pesticides in agriculture is not a new technology, since inorganic chemicals, such as arsenic or sulphur, were used for pest control in the nineteenth century. However it was only about 40 years ago, that synthetic organic chemicals were used as pesticides.

Since their introduction, the number and variety of pesticides has increased and the major groups used currently are shown in Table 1. From this table it can readily be seen that herbicides and not insecticides constitute the most commonly used type worldwide.

TABLE 1. PESTICIDE GROUPS

GROUP	% USED WORLDWIDE (US DOLLARS OF PRODUCT SOLD)
HERBICIDES	44
INSECTICIDES	31
FUNGICIDES	17
RODENTICIDES)	
)	
MISCELLANEOUS)	8
(PLANT GROWTH REGULATOR : BIOCIDES))	

(Source - Wood-MacKenzie)

Although concern is sometimes expressed about their safety to man and his environment, it must be recognised that potential risk cannot be totally eliminated and therefore the risk must be considered in the context of benefit. Pesticides undoubtedly do confer benefits on society and some of these are summarised in Table 2. However in countries where food is plentiful and malnutrition rare, it may be difficult to understand the benefit of improved food production, but pesticides do help to maintain this situation with good quality food. In other countries where malnutrition or even starvation may be more common, then increased food production is clearly necessary, but it must be linked with improved food distribution.

NATO ASI Series, Vol. H13
Toxicology of Pesticides: Experimental, Clinical and Regulatory Aspects. Edited by L. G. Costa et al.
© Springer-Verlag Berlin Heidelberg 1987

TABLE 2. BENEFITS OF PESTICIDES

IMPROVED FOOD PRODUCTION	- QUALITY AND QUANTITY
PUBLIC HEALTH	- VECTOR-BORNE DISEASE CONTROL - FLY CONTROL - PEST CONTROL
NON-EDIBLE CROP PROTECTION	- RUBBER : OIL-PALM : COTTON
DOMESTIC	- WOODWORM CONTROL - FLY CONTROL - AMATEUR GARDENER PRODUCTS - RODENTICIDES

Nevertheless public concern over pesticide use has led to increased demand by governments or manufacturers to undertake more studies evaluating human and environmental safety. This in turn has resulted in a more complex development process for producing pesticide products.

The aim of this paper is to describe pesticide safety assessment and its relationship to product development. The paper only concerns human safety, in particular safety to the agricultural worker. It does not attempt to discuss safety to the formulation or factory worker, nor does it attempt to describe environmental safety assessment, including the potential risk of consumption of foodstuff, which may contain pesticide residue. Through a clear understanding of the potential risk to the agricultural worker, measures to ensure safe use can be formulated and appropriate recommendations communicated to the customer. The type of measures for safe use and their communication will also be discussed.

PRODUCT DEVELOPMENT

This requires a multidisciplinary approach, shown in Figure 1. Besides the
need for chemists, biologists, statisticians and ecologists, toxicologists,
including medical advisers, also have an important role to play in product
development.

FIGURE 1

DEVELOPMENT OF A PESTICIDE

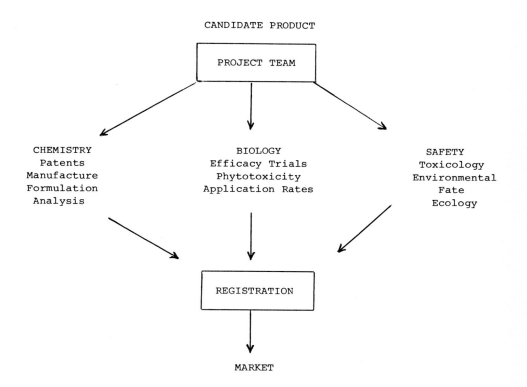

The development process involves a number of stages (Figure 2), beginning
with glasshouse screening, which normally involves between 10,000 and
20,000 chemicals per annum. The sources of the chemicals maybe from random
selection to purpose synthesis using biologically active analogues as
templates. Time taken to develop a pesticide from screening to production
of the final formulated product varies, but is usually of the order of 7 to
9 years.

FIGURE 2

DEVELOPMENT OF A PESTICIDE
– PHASE PROGRESSION

RESEARCH)	GLASSHOUSE SCREENING	-10-20,000	↑
PHASE)			
)	FIELD SCREENING	100-200	
	EVALUATION PHASE	~5-10	7-10 YEARS
	DEVELOPMENT PHASE	~1-2	
	MARKET	~ -1	↓

Of this large number of screened chemicals, only about 0.5 to 1% will show any useful biological activity, the characteristics of which must be more fully understood. In order to do this, these active chemicals are then put through field screening, when limited amounts of chemical are applied to small plots under different climatic and ecological conditions. This type of screening provides a better understanding of the spectrum of activity and examines the potential for adverse ecological effects more closely. After these two stages have been completed, the number of potential chemicals for development into products represents about 1 in 2000 of the total number of chemicals initially screened.

These two stages can be referred to as the 'Research Phase'. Those chemicals remaining now pass into the more detailed 'Evaluation Phase', when more complex field testing is undertaken. After this phase a decision must be taken as to which chemicals are suitable for product development and once this decision has been made, the chemical(s) now progresses into the 'Development Phase', which accounts for only 20% of those chemicals from the 'Evaluation Phase'. The number of final products emerging from this process will be approximately 1 per annum; half of those chemicals considered suitable for development.

The whole process is not without significant cost. Currently to develop a pesticide product having already invented the active ingredient, the cost is approximately 20-40 million pounds sterling. To this has to be added the cost of those chemicals, which were screened and found to be unacceptable for further development. The aim of this section was to give only a brief summary of product development, but the process has been reviewed more thoroughly by Lever (1985).

TOXICOLOGICAL ASSESSMENT

The requirement of a company to assess the toxicity of a chemical is necessary for two reasons :

(i)   To protect those members of staff involved in testing the chemical throughout the research and development phases. Not all these personnel need necessarily be part of the company developing the product since many companies require assistance from government, academic or commercial organisations during product development.

(ii) To protect members of the public, agricultural workers, consumers of treated produce and employees.

It must be stressed that the toxicological assessment only provides information on the potential for a chemical to cause harm or injury. The information on toxicity must be put into context of the use of the chemical and the frequency, dose and duration of exposure are important factors which need to be considered in the overall safety assessment.

Since it is totally impractical within the time available to test the toxicity of all 10 - 20,000 chemicals screened initially, the laboratory staff protection is done on 'worst-case' toxicity assumptions. To safeguard the staff therefore, chemical exposure is avoided during this screening stage using automated equipment wherever possible and full skin, eye and respiratory protective equipment is used to minimise exposure, when manual adjustment is necessary. When the screening stage has been completed a limited amount of toxicity studies (Table 3) is done on the 100 or so chemicals for field testing. Results of the tests done at this stage only give a crude indication of the chemicals' toxicity, but provide sufficient information to enable advice to be given on field worker protection.

After the field screening stage the number of chemicals to be developed is reduced to a more manageable size, for example 5-10 per annum and it is now practical to do more sophisticated toxicology. Much of this more sophisticated toxicology takes time, but it is advantageous to understand more about the chemicals' toxicity as early as possible for two reasons :

TABLE 3. ACUTE TOXICITY SCREENING TESTS

TEST	RELEVANCE
Acute Oral LD50	Swallowing - Accidental - Deliberate
Acute Dermal LD50	Handling Concentrated Product - Dermal Contact
Acute Inhalational LC50	Handling Volatile/Dusty Powders

Irritancy - Skin
        Eye
	Handling Concentrated Product
	        - Skin Or Eye Contact

Sensitisation Study	All Aspects Of Product Application

1.　　As development continues, chemicals are released into more and more diverse environments. Consequently the control a company may have over the safe use of its chemicals becomes less certain. It is therefore important that worker protection is based on release of less toxic formulations than to rely on the use of protective equipment and clothing for safety.

2.　　By predicting problems of toxicity early, it is possible to reduce development costs in some cases by stopping further development for particularly toxic chemicals and in others, through formulation or packaging changes, or by improved planning of further toxicity tests.

The type of toxicity results, which are very pertinent to the above are mutagenicity, carcinogenicity and teratogenicity. Unfortunately life-time feeding studies in animal models to assess carcinogenicity and teratogenicity are time consuming and an early answer is therefore not possible. For this reason a package of short term tests can be conducted prior to the development phase. This group of tests is referred to within ICI Plant Protection Division as a Research Toxicology Package and consists of :

1. Mutagenicity tests - in vitro and in vivo.

2. Modified teratology study - 'Chernoff Test'.

3. Subacute feeding study (rodent).

4. Special studies depending on available data.

Although reservations are often expressed about the relevance, for human safety, of some of the tests, for example mutagenicity tests, they do provide a practical solution to the problem of early assessment of the chemical's toxicity.

The results of this type of information can then be used to determine :
1. Whether further release of chemical should take place.

2. The likely acceptability for safe use as a final product.

3. Likely label recommendations for safe use.

Assuming that the chemical has no serious toxicity problems and that it is considered suitable for development, then the extent of toxicity testing increases in complexity and quantity. Table 4 summarises the type and probable relevance of the minimum number of toxicity tests, which are currently done during product development for registration. Many studies, for example the life-time feeding studies need to be planned carefully, so that preliminary ranging studies need to be done prior to starting them. Also some studies may not be completed to the satisfaction of either the laboratory or government authorities and so may need to be repeated.

TABLE 4.   REGULATORY TOXICOLOGY STUDIES

TEST	RELEVANCE	
	CONSUMER	AGRICULTURAL WORKER
SUBACUTE ORAL/DERMAL	✓	✓
LIFETIME FEEDING STUDY	✓	?
TERATOLOGY STUDY	✓	✓
REPRODUCTION STUDY	✓	?/✓
DELAYED NEUROTOXICITY STUDY	✓	✓
MUTAGENICITY STUDIES - In Vivo	✓	?
- In Vitro	✓	?

The overall conclusions from these studies is to enable a judgement to be made on the no observable effect level (NOEL), from which safety standards can be derived. The type of toxicological effect and the conditions under which it can be produced, including the frequency, duration and route of exposure are also important conclusions which need to be drawn.

SETTING THE SAFETY LIMITS

Although there are examples, where the relevance of animal data to human effects is understood, for the vast majority of industrial, household and agricultural chemicals it is not known. Consequently an assumption is made that man is always more sensitive to the chemical's effects by a factor (safety margin), which varies according to the type of toxicological effect observed. For example a serious irreversible effect such as a teratogenic event would probably demand a much higher safety margin than would a less serious or reversible effect.

By applying the safety margin to the NOEL, the upper limit of safe exposure to man can be determined. In the context of pesticides, the upper limit determined for the public consumption of foodstuff, which may contain pesticide residue, is termed the Acceptable Daily Intake (ADI) and is established by the World Health Organization.

In the context of exposure to industrial chemicals, safety limits are set for workers which are termed Threshold Limit Values (TLV). These TLVs assume that exposure to the chemical is predominantly via the inhalational route. Unfortunately no such standard is available for the agricultural worker, who may be exposed to pesticide and neither the ADI nor the TLV are appropriate. The ADI is inappropriate for at least two reasons :

1.   It is derived from the lowest NOEL, which is usually determined from studies involving continuous daily administration of a chemical to an animal over its life-time. Occupational exposure of agricultural workers to pesticides is rarely long-term and never throughout their life-span.

2.   The majority of toxicity studies used to determine the ADI involve oral administration. For the majority of pesticide applications most exposure is to the skin (British Agrochemicals Association - 1983 : Wolfe et al -1967 and 1974 Chester G and Woollen B H - 1982). Inhalational exposure usually represents a very small proportion of total exposure and in many instances is non-detectable (British Agrochemicals Association - 1983).

This second reason also explains why the TLV is also an inappropriate standard to use for the agricultural worker. There is therefore clearly a need to establish a standard for assessing user safety for agricultural workers and the next section deals with the principles by which this assessment can be made.

USER SAFETY ASSESSMENT

Ideally, to measure the safety of pesticide use, some sensitive marker of
acute toxicity could be measured before, during and after work. If the
work practice resulted in no abnormality of this marker and the marker
represented the most sensitive toxic effect, then the use of that product
could be assumed to be safe. For many years, this method has existed for
assessing organophosphate insecticide exposure by using serum or red-cell
cholinesterase as a marker.

Unfortunately, for the vast majority of new pesticides, this technique
does not exist. Instead assessments of exposure to skin and the respiratory
tract can be done and an estimate made of the amount absorbed. As
mentioned in the previous section, exposure to pesticides is predominantly
to the skin, so that if the dermal penetration of the pesticide can be
determined, an estimate can be made of the likely dose absorbed. Several
methods are available ranging from in vivo animal studies to laboratory
studies, which assess the barrier function of the outer skin layer the
stratum corneum of the epidermis. These methods to measure dermal
penetration have recently been reviewed by Dugard and Scott (1984). The
estimated absorbed dose can then be compared with an appropriate no-effect
level from animal studies and a margin of safety calculated.

Sometimes dermal exposure data is used as a basis for estimating absorbed
dose. The amount of pesticide absorbed through skin is measured using
controlled laboratory studies using animal models or even human volunteers
(Feldman and Maiback - 1974 : Wester R C et al - 1985) and the results can
be expressed as a percentage of the dose of chemical applied to the skin.
Using the percent absorbed from controlled laboratory studies, the amount
of chemical absorbed during pesticide application is calculated by
multiplying the percent absorbed by exposure. For example during use a
worker may be exposed to 100 mg per day and the laboratory study shows that
1% of the chemical is absorbed through skin from that formulation. The
amount absorbed can therefore be estimated to 1 mg/day.

This approach is clearly very simplistic and assumes that there is a fixed
relationship between dermal exposure and the amount of chemical absorbed
dermally : a relationship which has been challenged by the author (Chester
and Hart - 1986). However in the absence of other data it does offer a
practical and easily managed method of assessment.

Alternatively, several studies have been done which assess the dose absorbed by measuring the amount of chemical or its metabolite excreted in the urine (Chester and Hart - 1986 : Chester G C et al - 1987). By collecting and analysing all urine for a sufficiently long period after exposure, it is possible to measure the total amount of pesticide or metabolite excreted. It is also imperative to understand human pharmacokinetics, metabolism and renal handling of the pesticide, in particular it is necessary to know what percentage of an absorbed dose is excreted in the urine as parent compound and/or metabolites.

This method offers a more direct approach to measure absorbed dose following pesticide application, but the need to understand human pharmacokinetics of the pesticide may pose ethical problems in establishing controlled volunteer studies. Nevertheless it is approach, which can remove many assumptions involved with other methods eg 'percent absorption' and should always be considered during product development.

Whichever method is used to assess safety, the conditions under which the assessment is made must be pertinent to the conditions under which the product will be used normally. Normal use is, of course, very difficult to define and a certain amount of flexibility of understanding is required in order to describe it. However one definition which could apply is the use for which the product is intended and which includes minor predictable deviations from the label recommendations. Such deviations would include leakage or spillage from sprayers, excessive spray drift or the absence of recommended protective clothing.

Normal use must be clearly distinguished from situations which involve gross and unpredictable deviations from label recommendations. These situations may be termed 'misuse', and examples of which would include application of a pesticide at several times its recommended rate or through a piece of equipment not designed for that particular product's application. In addition to 'normal use' and 'misuse', situations may arise when a pesticide is used for a purpose for which it was never intended, for example, ingestion of a pesticide or its deliberate dermal application. These are obviously forms of abuse.

In summary the safety of pesticides to man must be considered in the context of their normal and proper use. Studies to assess this must therefore take into account conditions pertinent to normal use, including

the type of clothing worn or in some cases the degree of protective
clothing worn. In addition, in the event that certain pesticides are
associated with abuse or misuse, then effort should be made to develop
preventative measures.

HUMAN DATA

Much of what has been discussed above concerns the extrapolation of animal
toxicology data to humans. However for some types of effect, for example
acute effects, monitoring for adverse effects on users can provide valuable
information to complement the animal toxicology data. This assessment of
adverse effects can take the form of controlled field trials,
epidemiological studies or simply the recording of effects reported by
users of a product on an ad hoc basis or from questionnaires.

With all these methods it is not often that anything other than acute
effects can be assessed. In agricultural practice it is usual that a
multiplicity of products are used in any one season. Thus if any adverse
effects are discovered from, for example an epidemiological study it is
extremely difficult to isolate the offending agent although negative
results from such a study can provide some reassurance. It is therefore
rare that epidemiological studies examining chronic effects on agricultural
workers can provide conclusive information on safety or prevention,
although Howard's study of Malaysian plantation workers using paraquat
(Howard J K et al - 1981) is an exception.

Controlled field trials can also provide useful information about effects
on humans, but they tend to be limited to short term exposure and the
results are only valid for the conditions under which the study was
undertaken. There are many examples of this type of study, but the
majority involve acutely toxic pesticides and ones, which have some
measurable toxic effects, for example organophosphate or carbamate
insecticides. However examples involving other types of pesticide do
exist, for example paraquat (Swann - 1969) or pyrethroids (Hart T B -
1982) where this type of monitoring has been used successfully.

Lastly, the ad hoc or questionnaire monitoring of users of products can
also be used to assess human safety. The main disadvantage of this
approach is that it relies on a subjective assessment of events by the user
or interviewer (if a questionnaire is used). Consequently difficulty can
arise in interpreting results particularly with regard to establishing a
cause and effect relationship.

Nevertheless a carefully designed questionnaire can provide valuable information about product safety, particularly relating to acute local skin or eye effects. It is also worth noting that in the author's experience the collection of negative data can yield as much, if not more, information as the positive effect data, as well as enabling a particular problem with pesticide poisoning to be put in perspective.

With this type of data, which can be collected during development from field trials staff, from medical records of manufacturing or formulation workers or from agricultural workers using a mature product, it is important to define the type of effects observed. For example, there is a great deal of difference between a pesticide poisoning and a pesticide incident. The former involves exposure to a pesticide and adverse effects, which can be local or, less often, systemic and which result from that exposure.

A 'pesticide incident', however, may be defined as a situation in which exposure to a pesticide may or may not have occurred and adverse effects may or may not be present, but if effects are present, no evidence of a cause/effect relationship exists. In any event the exposure or the adverse effect cause concern, so that an enquiry is made but in general the 'incidents' will be of a non-serious nature.

POISON CONTROL CENTRES

The methods to collect human data as described in the previous section can all be done by individual companies. However the presence of poison control centres or poison information centres can help companies in providing information on pesticide poisonings or incidents.

The function of a poison control centre can vary depending principally on the country in which it exists. Consequently the type of information and service a poison control centre provides will depend very much on how the centre defines its function. In the United Kingdom, for example, the National Poison Control Centre, the functions of which are summarised in Table 5 below, is only available to medical staff and related professionals. Since the more serious poisonings will be referred to doctors and hospitals, this type of service will tend to detect the more serious cases. Other centres may be available to the public and consequently may have less bias towards serious cases in the collection of data.

TABLE 5.   FUNCTIONS OF A POISON CONTROL CENTRE

PROVISION OF INFORMATION RELEVANT TO DIAGNOSIS AND MANAGEMENT OF ACUTE POISONING IN MAN.

CONSULTATIVE CLINICAL SERVICE FOR TREATMENT OF POISONED PATIENTS.

LABORATORY SERVICE FOR ANALYSIS OF DRUGS AND POISONS.

EPIDEMIOLOGICAL STUDIES - ('TOXICOVIGILANCE')

TRAINING AND EDUCATION OF DOCTORS AND OTHER HEALTH SERVICE PERSONNEL.

PARTICIPATION IN MEASURES FOR PREVENTION OF POISONING INCLUDING EDUCATION OF THE PUBLIC.

(Source - the National Poisons Information Service, London)

These centres do, however, provide a centralised national collection system, but their main disadvantage is that there is not enough of them. It is true that they tend to be well-developed in Europe, the United States of America and isolated countries elsewhere. Unfortunately the majority of countries have only a rudimentary service or none at all. Without these centres it is virtually impossible to collect national data about pesticide poisoning, let alone to put such data in perspective in the context of other poisonings. Consequently for many countries there is no reliable data on pesticide poisonings or incidents and to try and estimate global poisoning can only be regarded as an educated guess.

PREVENTION OF PESTICIDE POISONING

If any preventative measure is to be effective, it must be directed at the circumstances by which poisoning with a particular pesticide, commonly occurs. For example, it is no use improving user safety with a particular product, if the main cause of poisoning is accidental ingestion. This means that a clear understanding is necessary of those situations which are likely to be hazardous. Information from field studies, epidemiology studies, user surveillance or poison control centres can help to define these situations.

The circumstances of poisoning and the route by which absorption commonly takes place are shown in Table 6 and preventative measures will be discussed in the context of these circumstances.

TABLE 6. PESTICIDE POISONING - ROUTE OF ABSORPTION

CIRCUMSTANCE	
WORK RELATED (Accidental)	- Dermal Inhalational Oral
ABUSE (Accidental or Deliberate)	- Oral Injection Dermal

1. Work Related Poisoning

Clearly ensuring the safe and effective use of a product must be the first priority of any company involved in the agrochemical business and it is usually achieved by advising users on how to apply the product safely. However it is important to ensure that the correct advice is given.

As mentioned above, the majority of pesticide applications result in skin, not inhalational exposure, so emphasis must be directed at protecting the skin from the pesticide. Whilst it is possible to recommend the use of extensive or elaborate protective clothing, this is often not practical due to cost and user inconvenience. It is totally impractical for example to wear a waterproof coverall when spraying in temperatures of 90-100°F with high humidity.

The first question to ask, therefore, is whether such protective clothing is really necessary and if it is, under what circumstances? Alternatively can skin protection be achieved by other means?

To help answer these questions, it is desirable, as with assessment of user safety, to understand the barrier function of the skin with regard to pesticide absorption. A number of techniques are now available to measure permeability, and these have been reviewed by Dugard and Scott -1984.

In general the concentrated pesticide formulations present a greater potential hazard than the diluted spray solutions, when this method of application is involved. The reason is that a more concentrated formulation is more acutely toxic and more irritant (to skin and eyes) than the dilute spray. Furthermore the ability of an active ingredient to penetrate skin from a concentrated solution will be better than from a diluted solution, especially in the presence of surfactants. We must therefore distinguish preventative measures for concentrate handling from those for spraying or applying the diluted product.

It may be necessary, from the above, to advise the use of protective equipment for handling concentrate only. This equipment might only need to be the use of protective gloves and/or eye-protection, both of which are practical to wear for short time periods, even in hot conditions. There is often no need to extend this equipment further, since several studies, including that by the British Agrochemicals Association (1983), have shown that hand exposure is the main route of exposure during mixing and loading.

For the remainder of the application, involving exposure to spray diluted formulations, only normal work clothing may be all that is required. However it must be recognised that the amount of chemical absorbed will depend on three factors :-

1.   Rate of skin penetration from formulation.
2.   Surface area of skin contact.
3.   Duration of skin contact.

The first two have been addressed through measurement and the use of protective clothing respectively, but if pesticide is allowed to remain on skin for a prolonged period of time, then more of the chemical can be absorbed. This, in most cases, can readily be prevented by simply washing the body and the work clothes regularly. Similarly splashes of concentrate should be removed from skin as soon as possible, together with affected clothing, which should be taken off and washed before re-use.

With some types of application, for example ultra-low volume spraying, more concentrated sprays are used than with conventional spraying. In these cases, careful consideration has to be given to dermal penetration of the chemical and the appropriate active ingredient or formulations chosen to ensure safe use. Protective equipment for many cases of spray application (not concentrate handling) should only be used as a last resort.

The importance of spray concentration and dermal penetration also means that the user must pay particular attention to the correct rate of application. Using products at rates, which are far higher than recommended can not only be counter productive in terms of biological efficacy, but can also be hazardous, because the spray concentration is increased.

In view of the above, typical recommended precautions for safe use
might be as shown in Table 7.

TABLE 7	SAFETY PRECAUTIONS
CAREFUL STORAGE	- Proper Labelled Container. - Do not decant.
CORRECT APPLICATION RATE	- Correct dilution of product.
CARE IN HANDLING CONCENTRATE	- Protective Equipment including glove, eye or face protection. - Washing.
AFTER WORK	- Wash. - Change and Wash work clothes regularly.
PROMPT FIRST AID TO SKIN AND EYES	

Face masks and respiratory protection have not been mentioned so far,
but there are instances when they might be necessary, for example
handling dusty solid formulations or using application equipment, such
as mistblowers, which produce a fine droplet spray. It is important
that the correct type of mask is used, which is both comfortable and
efficient in protection. There is a temptation to recommend face masks
for all types of spraying , but in the majority of cases, it is not
necessary and can be harmful. For example, since wearing a face mask
for a long period of time can be uncomfortable, manual adjustments are
common with the result that pesticide on the hands, contaminates the
mask and can lead to irritation around the nose and mouth.

Lastly safe practice for pesticide application should always include
regular sprayer maintainance and non-overfilling of sprayers. Leakages
and spillages from sprayers represent a major and sometimes serious
source of contamination particularly when the tank is in close
proximity to the body. It is a very desirable aim to have well-
maintained sprayers used, but often cost and lack of care prevent this
being achieved.

2. Accidental Poisoning

In order to distinguish from work related poisoning, which can be accidental, this type of poisoning involves, almost invariably, ingestion of chemical. The most common circumstances surrounding accidental poisoning involve the mistaken identity of a pesticide formulation for a common beverage or the accidental ingestion by children. This situation is more likely to occur if storage of chemical is careless and/or in the incorrect container and if the chemical formulation resembles a common beverage, at least in colour.

Pesticides should be always stored properly, preferably under lock and key, in their original container, correctly labelled. The practice of decanting into alternative containers, for example beer or lemonade bottles, should be actively discouraged. Worse still, in some countries pesticides have been sold in modified drinks containers, such as rum or wine bottles, often incorrectly labelled. This practice represents one of crass stupidity and should be stopped.

Pesticide manufacturers can do a great deal to prevent accidental poisoning, by not only using the most appropriate package or bottle, but also by formulation. For example some of the more toxic pesticides have undergone formulation changes with addition of colouring agents (organophosphates, paraquat), or odoured chemicals (paraquat) to help prevent mistaken identity. Dilution of a product may also be used to help prevent accidental poisoning. This, in fact, the basis of the application of the Poisons Regulations in the United Kingdom, in that those formulations suitable for use by the general public are much more dilute than their couterparts used in Agriculture by a more restricted and professional population.

National Governments may also help in prevention of accidental pesticide poisoning in a number of ways. Through the regulatory scheme, they can ensure safety standards for packaging and formulation are uniform. It is no use one manufacturer applying safety standards to a particular product, when others may ignore them for the same product.

Governments can assist with restricting the sales particularly of the more toxic chemicals, to bone fida users and not allowing the public to buy such products. In some countries, for example the United Kingdom and the United States of America such a scheme operates well.

Unfortunately in many countries, the necessary infrastructure for maintaining a restrictive sales system does not exist and therefore, although law can be passed, it is not enforceable.

It is not possible to erradicate accidents completely, but many can and should be prevented. Measures such as those described above can play a major role in achieving that objective, particularly the careful storage of pesticide products. For example a firearm left lying around could cause an accident, but if locked away the risk is very much reduced.

3. Deliberate Abuse

The use of synthetic chemicals for suicide or homicide has unfortunately become a modern-day scourge. In this respect pesticides are no different from other types of chemical, such as drugs or household chemicals and examples, such as arsenate, organophosphate insecticides and paraquat are fairly well-known.

There are several major differences between suicidal and accidental poisoning, which explain why suicidal poisoning is so difficult to prevent. In accidental poisoning, the volume dose of chemical swallowed is almost always small, but the dose swallowed by suicide is variable but often large. Thus dilution of a product may be effective in preventing accidents, but success for suicide prevention is very doubtful, unless the dilution factor is large.

Suicides often delay admission to hospital and may not admit swallowing chemical, neither of which is true for accidents. Consequently treatment of suicide cases is often delayed, whereas accidental poisonings can be treated quickly.

Preventative measures for accidental poisonings, which involve making them aware of the toxic nature of the product, for example label warnings, formulation changes or publicity, can be effective, but with suicides can do the reverse and stimulate further attempts. This has been shown to be true for paraquat poisonings (Hart and Bramley - 1983).

Figure 3 is a graph of United Kingdom paraquat poisonings from 1965 until 1984. The rise of poisonings, due to suicide, between 1970 and 1975 is likely to be due to the immense publicity given to the product's toxicity between those years. Availability of product did

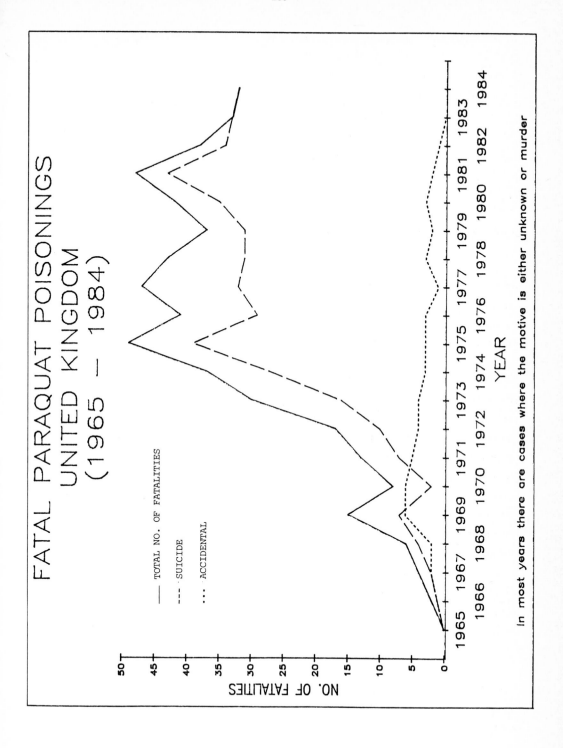

FATAL PARAQUAT POISONINGS
UNITED KINGDOM
(1965 — 1984)

TOTAL NO. OF FATALITIES
SUICIDE
ACCIDENTAL

NO. OF FATALITIES

YEAR

In most years there are cases where the motive is either unknown or murder

not increase, nor were suicides more common during or around those years and so neither of these factors can account for the rise. This same phenomenon has been seen for other types of suicide from publication of coroner's reports (Surtees S T -1982 : Barraclough B M et al 1977).

It must be stressed, though, that publicity is not likely to influence the total number of suicides, but only the method of choice.

In summary, it is very difficult to prevent deliberate abuse, but can be easy to stimulate. Perhaps a more responsible attitude by the media in avoiding mention of specific suicide methods and research into the basic motivation of suicide is likely to be a more profitable route for prevention than removing the means of committing suicide.

COMMUNICATION OF SAFE PRACTICE

Once the appropriate messages on safe practice have been formulated, communication of them must not be neglected. The product label is probably the most important means of communication a manufacturer may have with the customer. It is important to ensure, therefore, that the messages are transmitted in a legible form, which is easily understood.

In any population, there is a proportion of people who can read, another who cannot read and one suspects a silent majority, who can read, but do not do so. For these last two groups, a great deal can be done to improve communication.

There is obviously a difficulty in writing extensive precautions or pictures on labels stuck onto relatively small bottles. To get around the space problem, some manufacturers have used pocket inserts on labels - 'lablets' or folding labels, examples of which are shown in Figures 4 and 5. At the same time, attempts have been made to improve legibility with the use of pictorial instructions - 'pictograms' (Figure 6). It remains to be seen how effective these changes will be, but commonsense leads one to believe they are improvements.

FIGURE 4
LABEL ENVELOPE – 'LABLET'

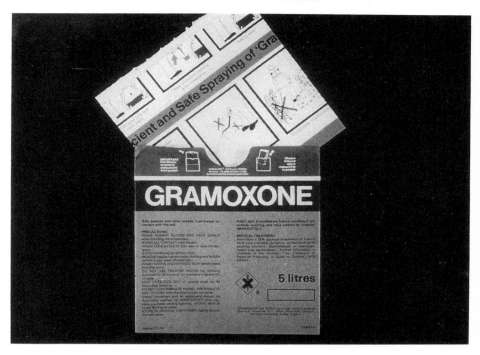

FIGURE 5
EXPANDING LABEL

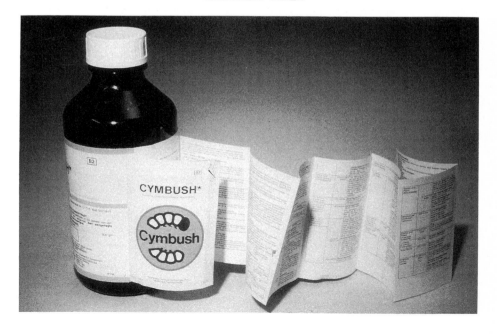

FIGURE 6
'PICTOGRAM' DEPICTING SAFE USE RECOMMENDATIONS

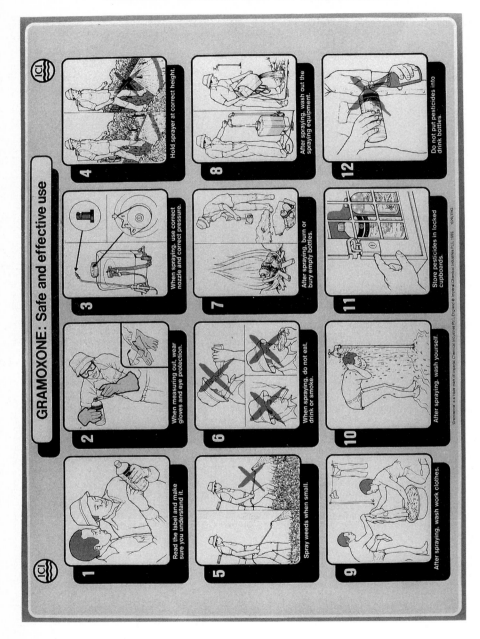

Several other routes of communication also exist for manufacturers or trade organisations. These include posters, booklets, audio-visual aids and the media, especially television and radio. Posters and booklet on safe and effective use have been produced by * 'GIFAP', the international pesticide trade organisation.

Lastly safety messages can be sent out as they stand or encorporated with other messages. For example safe use can be encorporated into messages about effective use of pesticides at farmer meetings or through advertising. It is likely that education of farmers to use pesticides safely and effectively is more likely to achieve the desired objective than campaigns to promote safety use only.

CONCLUSION

The diversity of pesticides and their uses, together with the increased demand by society for safety testing has resulted in a rather complex development process for a pesticide product. This paper briefly describes that process, with particular emphasis on how safety testing is done and at what stage of development. The paper confines itself to human safety evaluation to the user, with some attention given to the misuse or abuse of pesticides by the general public.

Attention is also given to methods used to monitor for acute effects on the user or general public. Several types of active and passive monitoring are described, in particular the role of the poison control centres in assessment of all forms of acute adverse effects, especially those arising from pesticide abuse. The lack of poison control centres makes it difficult to understand the extent and nature of pesticide poisoning in many countries, but in those countries where they are present the type of pesticide poisoning and its perspective with other poisoning can be defined.

* Groupment International des Associations Nationales de Fabricants de Products   Agrochimiques. (International Group of National Associations of Manufacturers   of Agrochemical Product).

Assessment of long term or chronic effects with epidemiology studies on agricultural workers has been conducted. Unfortunately due to the variety and number of products used by a worker during any one season, it is usually difficult to draw any firm conclusions from such studies.

However from a good understanding of circumstances under which pesticides can be used safely and the circumstances under which they may be abused, it is possible to make appropriate recommendations for safe use and prevention of misuse and abuse. Measures, which can be taken by governments and industry to prevent poisoning are discussed and the effect the media may have on pesticide poisoning, particularly suicide is described.

There is no evidence at present to suggest that pesticide poisoning from 'normal use' is either common or serious, but when poisoning does occur it is usually associated with abuse or misuse. The good safety record for 'normal use' of pesticides can and should be maintained; the methods described in this paper can help to achieve that objective. Accidental poisoning can also be prevented in many cases through careful storage, packaging, formulation and labelling, but unfortunately these methods are unlikely to influence deliberate poisonings. It is worth repeating that well-used phrase, 'there is no such thing as a safe chemical only safe ways of using them'.

Barraclough, B.M. et al (1977)
   Do newspaper reports of coroner's inquests incite people to commit
   suicide?
   B.J. Psych. 131 528-532.
British Agrochemicals Association (1983)
   Spray Operator Safety Study.
Chester, G. et al (1987)
   Worker Exposure to and Absorption of Cypermethrin During Aerial
   Application of an 'Ultra Low Volume' Formulation to Cotton.
   Arch. Environ. Contam. Toxicol. - 16 69-78.
Chester, G. and Hart, T.B. (1986)
   Biological Monitoring of a Herbicide Applied Through Backpack and
   Vehicle Sprayers.
   Tox. Letters - 33 137-149.
Chester, G. and Woollen, B.H. (1982)
   A Study of the Occupational Exposure of Malaysian Plantation Workers
   to Paraquat.
   Brit, J. Ind. Medicine 38 23-33.
Dugard, P.H. and Scott, R.C. (1984)
   Absorption through skin.
   Internat. Encyclopedia of Pharmacology and Therepeutics Section 10
   The Chemotherapy of Psoriasis Chapter 8 pp 125-144.
   Edited by H.P. Baden - Pergammon Press, Oxford.
Feldman, R.J. and Maiback, H. (1974)
   Percutaneous Penetration of some Pesticides and Herbicides in Man.
   Toxicol. and Appl. Pharmacol. 28 126-132.
Hart, T.B. et al (1982)
   A Study of the Exposure and Health of Indian Workers Spraying 'Ambush'
   and 'Cymbush' on Cotton using High Volume Hand-held Spray Applicators.
   ICI Internal Report.
Hart, T.B. and Bramley, A. (1983)
   Paraquat Poisoning in the United Kingdom.
   Human Toxicol, 2 417.
Howard, J.K. Sabapathy, N.N. and Whitehead, A.P. (1981)
   A Study of the Health of Malaysian Plantation Workers with particular
   reference to Paraquat Sprayment.
   Brit. J. Ind. Medicine 38 110-116.
Lever, B.G. (1985)
   The Economics of PGR Research - Risk or Reward.
   British Plant Growth Regulator Group - Monograph 13 - 1985.
Surtees, S.J. (1982)
   Suicide and Accidental Death at Beachy Head.
   B.M.J. - 284 321-324.
Swan, A.A.B (1969)
   Exposure of Spray Workers to Paraquat.
   Brit. J. Ind. Medicine 26 322-329.
Wester R.C. et al (1985)
   In Vitro Percutaneous Absorption of Paraquat from Hand, Leg and Forearm .
   J.Toxicol. Environ. Health 14 (5-6) 759-762.
Wolfe et al (1967)
   Exposure of Workers to Pesticides.
   Arch. Environ. Health. 14 622-6.
Wolfe et al (1974)
   Exposure of Mosquito Control Workers to Fenthion.
   Mosquito news 34 (No 3) 263-267.

Role of Biotechnology in Pesticide Development: *Bacillus thuringiensis* as an Example

David A. Fischhoff

Biological Sciences

Monsanto Co.

700 Chesterfield Village Parkway

Chesterfield, MO 63198

USA

Introduction

Bacillus thuringiensis (*B.t.*) is a spore forming bacterium which produces an insecticidal parasporal crystal upon sporulation. The insecticidal nature of *B.t.* has been known for several decades. For many years *B.t.* has served as the basis of successful biological insecticides such as Dipel (Abbot) and Thuricide (Sandoz). To produce these insecticides *B.t.* is fermented until spores and crystals are produced. The mixture of spores and crystals is then formulated to allow effective application to crop plants. Products such as Dipel have been used effectively on more than 50 species of Lepidopteran pests on over 200 crops (Wilcox *et al.*, 1986).

Two features of *B.t.* have made it a popular and useful insecticide. First, *B.t.* is considered extremely safe. According to the Farm Chemicals Handbook (1985) *B.t.* is "harmless to humans, animals and useful insects." Second, *B.t.* is a highly specific insecticide; most strains of *B.t.* show toxicity to only a single order of insects (e.g., Lepidoptera) as described below. These features have also made *B.t.* an attractive target for biotechnology.

The current *B.t.* products are an example of classical industrial microbiology. They have been created through traditional microbiological practices such as strain isolation and improvement and fermentation optimization. Recently, the tools of modern biotechnology such as gene cloning and DNA sequencing have begun to be applied to *B.t.* Through biotechnology it is likely that we can create novel pesticidal products based on *B.t.*

NATO ASI Series, Vol. H13
Toxicology of Pesticides: Experimental, Clinical
and Regulatory Aspects. Edited by L. G. Costa et al.
© Springer-Verlag Berlin Heidelberg 1987

In this report I summarize the current status of biotechnology as applied to *B.t.* with an emphasis on recent results in molecular biology.

Varieties of *B.t.*

There are hundreds if not thousands of isolates of *B.t.* which have been categorized. The vast majority of isolates which have been tested have shown activity only against larvae of Lepidopteran insects such as *Manduca sexta*, *Heliothis virescens* and *Trichoplusia ni*. Among the Lepidopterans, not all are equally sensitive to *B.t.* For example, Spodoptera species tend to be relatively insensitive.

The Lepidopteran specific *B.t.* strains have been categorized according to flagellar serotype and crystal serotype as well as activity spectrum against various insects (Dulmage, 1981). Among the better studied varieties of Leptidopteran specific *B.t.* are *B.t. kurstaki* HD1 which is the strain used in Dipel, *B.t. kurstaki* HD73, *B.t. dendrolimus*, *B.t. sotto* and *B.t. berliner*.

In recent years two types of *B.t.* with new insecticidal specificities have been discovered. *B.t. israelensis* is toxic to larvae of several Dipteran species (mosquitoes and black flies) but not to Lepidopteran larvae (Goldberg and Margalit, 1977). More recently, two Coleopteran specific strains, *B.t. tenebrionis* and *B.t. san diego*, have been described (Krieg et al., 1983; Krieg *et al.*, 1984; Herrnstadt *et al.*, 1986). These strains have shown activity against Colorado potato beetle and other Coleopteran pests.

B.t. Toxins

The insecticidal activity of *B.t.* resides in the parasporal crystal (Angus, 1954). Intact crystals can be isolated from sporulated cultures of *B.t.* by density gradient centrifugation, and these isolated crystals have been characterized for all three types of *B.t.*

Crystals of the Lepidopteran specific *B.t. kurstaki* are composed of protein subunits of approximately 130,000 daltons (Bulla *et al.*, 1977). These subunits can be released from the crystal by solubilization at

alkaline pH in the presence of reducing agents. In some strains such as
B.t. kurstaki HD73 there appears to be a single protein subunit, while in
other strains such as *B.t. kurstaki* HD1 there appear to be two or more
very similar proteins in the crystal (Wilcox *et al.*, 1986). The 130,000
dalton protein is considered to be a protoxin because it is toxic to
larvae only after ingestion but not after injection (Angus, 1956). The
protoxin can be converted to the active toxin by digestion with proteases.
These proteases can be either gut proteases from sensitive larvae or
purified proteases such as trypsin. Protease digested protoxin is lethal
by injection or ingestion (for review see Fast, 1981). The 130,000 dalton
protein is also capable of killing some Lepidopteran cell lines *in vitro*
but again only after protease activation (Murphy *et al.*, 1976; Johnson
1981).

It has been possible to isolate proteolytic fragments of the protoxin
which retain full toxic activity. Bulla *et al.* (1981) found that a toxin
fragment of 68,000 daltons could be derived from *B.t. kurstaki* crystals
upon prolonged incubation. Nagamatsu *et al.* (1984) found that a fragment
of 58,000 daltons with full toxicity could be isolated from trypsin
digested solubilized protoxin of *B.t. dendrolimus*.

Based on the above observations a reasonable scheme for the action of
Lepidopteran specific *B.t.* toxin is as follows. Larvae ingest the whole
crystal which is solubilized by the alkaline pH of the larval midgut to
release the protoxin subunits. These protoxin subunits are then digested
to the active toxin by the action of gut proteases. Finally, the toxin
acts on the susceptible gut cells.

Crystals of the Coleopteran specific *B.t.* strains also appear to be
composed of a single protein subunit, but of a much smaller size than the
Lepidopteran protoxin. Bernhard (1986) isolated a 68,000 dalton protein
from crystals of *B.t.t.* and Herrnstadt *et al.* (1986) have observed a
64,000 dalton protein from *B.t. san diego* crystals. These isolated
proteins are toxic upon ingestion by sensitive Coleopterans; however, it
is not yet known if these proteins represent protoxins or active toxins

The insecticidal crystals of *B.t. israelensis* are apparently more complex.
There are apparently at least two and perhaps up to four proteins in these

crystals which might have specific mosquitocidal activity with molecular
weights ranging from 230,000 daltons to 28,000 daltons (Hurley *et al.*,
1985; Visser *et al.*, 1986; Sekar, 1986). The 28,000 dalton protein is
most likely a cytolytic factor (Visser *et al.*, 1986) but might also have
some insecticidal activity (Ward *et al.*, 1986). Several groups have now
reported cloning genes corresponding to various proteins from *B.t. israel-*
ensis crystals (see below), and these cloned genes will help to clarify
this complex situation.

Mode of Action of *B.t.* Toxin

The mode of action of the Lepidopteran specific *B.t.* toxin has been
studied at both the histopathological and molecular levels. Within
minutes after ingestion of crystals, sensitive larvae cease feeding. A
general paralysis follows. At the histopathological level, the cells of
the midgut epithelium have been shown to swell and eventually burst after
toxin treatment (Luthy and Ebersold, 1981). Similar effects are seen with
cultured cell lines *in vitro* (Murphy *et al.*, 1976). At the molecular
level, the toxin has been shown to have effects on ion transport in both
midgut cells and cultured cell lines (Harvey and Wolfsberger, 1979; Griego
et al., 1979; English and Cantley, 1985; Himeno *et al.*, 1985). Although
these effects occur rapidly, it is not known what the primary target for
the toxin is. There is likely to be a specific receptor for the toxin on
the surface of the sensitive cells and a recent report indicates that this
receptor might be a high molecular weight glycoprotein (Knowles and Ellar,
1986).

Molecular Cloning of *B.t.* Toxin Genes

Much current work is directed at the isolation and characterization of
genes encoding *B.t.* toxins. The analysis of such cloned genes has already
yielded important insights into toxin structure and function.

Several groups have reported cloning genes for Lepidopteran specific
toxins. Most of these genes have been cloned in *E. coli* either utilizing
antibodies to purified toxin to detect expression of the toxin or utilizing synthetic oligonucleotide probes based on the toxin amino acid
sequence to detect toxin genes by hybridization. The genes cloned include
several from *B.t. kurstaki* HD1 (Schnepf and Whiteley, 1981; Held *et al.*,

1982; Watrud *et al.*, 1985; Shivakumar *et al.*, 1986; Thorne *et al.*, 1986), and genes from *B.t. kurstaki* HD73 (Adang *et al.*, 1985), *B.t. sotto* (Shibano *et al.*, 1985), *B.t. berliner* (Klier *et al.*, 1982; Wabiko *et al.*, 1986), *B.t. aizawa* (Klier *et al.*, 1985) and *B.t. thuringiensis* (Honigman *et al.*, 1986). In general these genes have been shown to express toxin protein in *E. coli*, and extracts of *E. coli* harboring these genes are toxic to Lepidopteran larvae.

The cloned toxin genes have been used as molecular probes to determine the toxin gene composition of many Lepidopteran active *B.t.* strains (Kronstad *et al.*, 1983). This analysis has shown that while some strains (e.g., *B.t. kurstaki* HD73) contain only a single toxin gene, many other strains contain multiple genes. *B.t. kurstaki* HD1 (the Dipel strain) has three distinct toxin genes (Wilcox *et al.*, 1986). Based on specific restriction endonuclease fragments associated with each gene, it appears that the toxin genes in many strains fall into common families (Kronstad and Whiteley, 1986).

DNA sequences and derived amino acid sequences of the toxin proteins have been determined for several of these genes. All of the genes encode proteins of between 1156 and 1178 amino acids which are largely homologous. In some cases genes isolated from strains which have been considered as distinct varieties have been found to be nearly identical. For example, the *B.t. kurstaki* HD1 gene sequenced by Schnepf *et al.* (1985) is nearly identical to the *B.t. sotto* gene of Shibano *et al.* (1985). Similarly, a *B.t. berliner* gene (Wabiko *et al.*, 1986) is nearly identical in sequence to another *B.t. kurstaki* HD1 gene (D.F., unpublished). On the other hand, a third gene from *B.t. kurstaki* HD1 (Thorne *et al.*, 1986) is clearly different in sequence from the two mentioned above.

Pairwise comparison of several toxin sequences has been performed (Wabiko *et al.*, 1986; D.F., unpublished) to localize the regions of similarity and difference. From the N-terminus the genes are all nearly identical for approximately the first 330 amino acids. Similarly, from about amino acid 600 through the C-terminus the genes are largely the same, although in this region there appear to be two classes of genes which differ from one another by an insertion of 26 amino acids as well as by several isolated amino acid differences. However, in their central regions the genes show much more variation. There is as much as 65% nonhomology in amino acid sequence between some genes from *B.t. kurstaki* HD1 and *B.t. kurstaki* HD73 between amino acids 350 and 600, for example. Based on published sequences there are at least four distinct types of Lepidopteran toxins which differ substantially in this central region. It is known that when compared at the level of whole strains (not cloned genes) HD73 is more effective against *H. virescens* than *T. ni* while HD1 is more effective against *T. ni* than *H. virescens* (Dulmage, 1981). It is likely that differences in the toxin protein sequences and the presence of multiple genes in some strains are responsible for this differential toxicity.

Because the cloned genes express active toxin in *E. coli*, it has been possible to construct deletion variants of the toxin genes and determine which segments of the gene are necessary for toxicity. This analysis has been carried out for the *B.t. kurstaki* HD73 gene (Adang *et al.*, 1985) the *B.t. sotto* gene (Shibano *et al.*, 1985) two genes from *B.t. kurstaki* HD1 (Schnepf and Whiteley, 1985; D.F., unpublished) and a *B.t. berliner* gene (Wabiko *et al.*, 1986). Results from all of these genes are largely the same. The portion of the gene necessary to encode a functional toxin seems to reside in the N-terminal half of the protein from about amino acid 10 through amino acid 615. Interestingly, this is approximately the same region of the protein which is released as an active toxin from trypsin treated protoxin *in vitro*. The C-terminal half of the molecule apparently is not necessary for toxicity.

Cloning of genes for Dipteran specific toxins from *B.t. israelensis* has also been reported. Waalwijk *et al.* (1985) and Ward and Ellar (1986) have cloned and sequenced the gene for the 28,000 dalton cytolytic factor, and Ward *et al.* (1986) report that the protein produced by the cloned gene when expressed in B. subtilis is mosquitocidal as well as cytolytic. Sekar and Carlton (1985) have cloned a separate gene from *B.t. israelensis* which has only mosquitocidal activity. The sequence of this gene has not been reported, but the gene apparently encodes a 130,000 dalton protein (Sekar, 1986). A third distinct gene with mosquitocidal but no cytolytic activity has been cloned and sequenced by Thorne *et al.* (1986); this clone produced a 58,000 dalton protein in B. subtilis. Computer analysis of this gene revealed several regions of homology with a Lepidopteran toxin gene from *B.t. kurstaki* HD1 (Thorne *et al.*, 1986).

Herrnstadt *et al.* (1986) have reported the cloning of a gene for the Coleopteran specific toxin from *B.t. san diego*. Extracts of *E. coli* harboring this gene were toxic to *P. luteola*. DNA sequence data has not yet been reported.

The gene cloning, expression in *E. coli* and DNA sequencing of toxin genes described above is the starting point for further molecular biological analysis and manipulation of *B.t.* toxin genes. Already the deletion analysis of Lepidopteran toxin genes has allowed the construction of much smaller genes with full toxicity. DNA sequence comparisons have revealed both conserved and nonconserved regions. It should prove possible to begin to further dissect the cloned genes and identify functional domains, for example, the regions required for receptor binding. Eventually we should be able to understand the basis for the high degree of specificity which these toxins show and perhaps to alter that specificity as well.

Genetically Engineered Microbial Pesticides

The isolation and characterization of *B.t.* toxin genes has also led to another biotechnological innovation, the genetically engineered microbial pesticide. In an engineered microbial pesticide, the *B.t.* gene would be cloned into an alternative microbial host which would have superior properties for field application. For example, the new microbial host might colonize specific plants or parts of plants (e.g., roots). Additionally, the new host, by virtue of multiplying along with the plant, might minimize the need for repeated application of the pesticide during the growing season.

Watrud *et al.* (1985) have put some of these ideas into practice and created a model genetically engineered microbial pesticide. They have isolated and characterized a strain of *Pseudomonas fluorescense* which is an efficient colonizer of the roots of corn plants. Using broad host range plasmid vectors capable of replicating in *Pseudomonads*, these workers have introduced a toxin gene from *B.t. kurstaki* HD1 into this corn root colonizer. The toxin gene is expressed in this host at pesticidal levels. In the growth chamber this engineered microbe survives on corn roots as well as the nonengineered parent.

In order to minimize the potential for genetic exchange of the *B.t.* toxin gene from the engineered *Pseudomonads* to other soil microbes, Obukowicz *et al.* (1986a) have utilized the transposable DNA element Tn5 to insert the toxin gene into the chromosome of several root colonizing microbes. In a further advance, these workers created disarmed Tn5 elements which could no longer promote DNA transposition and used these to integrate the toxin gene into the chromosome of root colonizing *Pseudomonads* (1986b). In this last case the B.t toxin gene should be as stable genetically as any other chromosomal gene.

Plant Genetic Engineering

The past few years have seen rapid advances in our ability to genetically engineer plants (for review see Jaworski *et al.*, 1986). The system that is most generally used is based on the Ti plasmid of *Agrobacterium tumefaciens*. This bacterium can naturally transfer a segment of the Ti DNA (called T-DNA) to plant cells causing crown gall disease. It has proved possible to "disarm" the T-DNA so that it no longer causes any disease symptoms in plants but still promotes DNA transfer to plant cells. Using disarmed Ti plasmids and intermediate plasmid vectors containing genes of interest it is possible to transfer desired foreign DNA into plant cells (Fraley *et al.*, 1985). Whole plants containing the introduced DNA in each of their cells can be regenerated from such transformed cells, and the introduced DNA is inherited by the progeny (Horsch *et al.*, 1984; Horsch *et al.*, 1985)

These plant genetic engineering systems have now been used to create plants with potentially useful agronomic properties. Petunia plants have been made resistant to the herbicide glyphosate through the overproduction of the target enzyme for the herbicide (Shah *et al.*, 1986). Tobacco plants showing resistance to tobacco mosaic virus have also been engineered through the expression of a gene for the viral coat protein (Abel *et al.*, 1986).

In a similar fashion it is possible to construct chimeric *B.t.* toxin genes engineered for expression in plant cells. By transforming plants with such genes we will soon be able to determine whether expression of *B. t.* toxin in plants is a feasible method for conferring resistance to insect pasts.

References

Abel PP, Nelson RS, De B, Hoffmann N, Rogers SG, Fraley RT, Beachy RN (1986) Delay of disease development in transgenic plants that express the tobacco mosaic virus coat protein gene. Science 232:738-743

Adang MJ, Staver MJ, Rochleau TA, Leighton J, Barker RF, Thompson DV (1985) Characterized full length and truncated clones of the crystal protein of *Bacillus thuringiensis* subsp. *kurstaki* HD73 and their toxicity to Manduca sexta. Gene 36:289-300

Angus TA (1954) A bacterial toxin paralysing silkworm larvae. Nature 173:545-546

Angus, TA (1956) Extraction, purification and properties of *Bacillus sotto* toxin. Can J Microbiol 2:416-426

Bernhard K (1986) Studies on the delta endotoxin of *Bacillus thuringiensis* var. *tenebrionis*. FEMS Microbiol Lett 33:261-265

Bulla LA, Kramer KJ, Cox DJ, Jones BL, Davidson LI, Lookhart GL (1981) Purification and characterization of the entomocidal protoxin of *Bacillus thuringiensis*. J Biol Chem 256:3000-3004

Bulla LA, Kramer KJ, Davidson LI (1977) Characterization of the entomocidal parasporal crystal of *Bacillus thuringiensis*. J Bacteriol 130:375-383

Dulmage HT (1981) Insecticidal activity of isolates of *Bacillus thuringiensis* and their potential for pest control, pp. 193-222. In Burges HD (ed) Microbial control of pests and plant diseases 1970-1980. Academic Press New York

English LH, Cantley LC (1985) Delta endotoxin inhibits Rb+ uptake, lowers cytoplasmic pH and inhibits a K+-ATPase in *Manduca sexta* CHE cells. J Memb Biol 85:199-204

Fast PG (1981) The crystal toxin of *Bacillus thuringiensis*, pp. 223-248. In Burges HD (ed) Microbial control of pests and plant diseases 1970-1980. Academic Press New York

Fraley RT, Rogers SG, Horsch RB, Eichholtz DA, Flick JS, Fink CL, Hoffmann NL, Sanders PR (1985) The SEV system: A new disarmed Ti plasmid vector system for plant transformation. Biotechnology 3:629-635

Goldberg LJ, Margalit J (1977) A bacterial spore demonstrating rapid larvicidal activity against Anopheles sergentii, Uranatenia unguiculata, Culex univittatus, Aedes aegypti and Culex pipiens. Mosquito News 37:353-358

Griego VM, Moffett D, Spence KD (1979) Inhibition of active K+ transport in the tobacco hornworm midgut after ingestion of *Bacillus thuringiensis* endotoxin. J Insect Physiol 25:283-288

Harvey WR, Wolfersberger MG (1979) Mechanism of inhibition of active potassium transport in isolated midgut of *Manduca sexta* by *Bacillus thuringiensis* endotoxin. J Exp Biol 83:293-304

Held GA, Bulla LA, Ferrari E, Hoch J, Aronson AI, Minnich SA (1982) Cloning and localization of the lepidopteran protoxin of *Bacillus thuringiensis* subsp. *kurstaki*. Proc Nat Acad Sci USA 79:6065-6069

Herrnstadt C, Soares GG, Wilcox ER, Edwards DL (1986) A new strain of *Bacillus* thuringiensis with activity against Coleopteran insects. Biotechnology 4:305-308

Himeno M, Koyama N, Funato T, Komano T (1985) Mechanism of action of *Bacillus thuringiensis* insecticidal delta endotoxin on insect cells *in vitro*. Agric Biol Chem 49:1461-1468

Honigman A, Nedjar-Pazerini G, Yawetz A, Oron U, Schuster S, Broza M,
    Sneh B (1986) Cloning and expression of the lepidopteran toxin
    produced by *Bacillus thuringiensis* var. *thuringiensis* in *E. coli* Gene
    42:69-77
Horsch RB, Fraley RT, Rogers SG, Sanders PR, Lloyd A, Hoffman N (1984)
    Inheritance of functional foreign genes in plants. Science
    223:496-498
Horsch RB, Fry JE, Hoffman NL, Eichholtz DA, Rogers SG, Fraley RT (1985)
    A simple and general method for transferring genes into plants.
    Science 227:1229-1231
Hurley JM, Lee SG, Andrews RE, Klowden MJ, Bulla LA (1985) Separation of
    the cytolytic and mosquitocidal proteins of *Bacillus thuringiensis*
    subsp. *israelensis*. Biochem Biophys Res Comm 126:961-965
Jaworski EG, Fraley RT, Rogers SG, Horsch RB, Beachy RN, Chua N-H (1986)
    Genetic transformation of plants, pp. 383-393. In Inouye M, Sarma R
    (eds) Protein Engineering. Academic Press New York
Johnson DE (1981) Toxicity of *Bacillus thuringiensis* entomocidal protein
    toward cultured insect tissue. J Invert Pathol 38:94-101
Klier A, Fargette F, Ribier J, Rapoport G (1982) Cloning and expression
    of the crystal protein genes from *Bacillus thuringiensis* strain
    strain *berliner* 1715. EMBO J 1:791-799
Klier A, Lereclus D, Ribier J, Bourgouin C, Menou G, Lecadet M, Rapoport
    G (1985) Cloning and expression in *E. coli* of the crystal protein
    gene from *Bacillus thuringiensis* strain aizawa 7-29 and comparison of
    the structural organization of genes from different serotypes, pp.
    217-224. In Hoch JA, Setlow P (eds) Molecular biology of microbial
    differentiation. American Society for Microbiology Washington
Knowles BH, Ellar DJ (1986) Characterization and partial purification of
    a plasma membrane receptor for *Bacillus thuringiensis* var. *kurstaki*
    Lepidopteran specific endotoxin. J Cell Sci 83:89-101
Krieg A, Huger AM, Langenbruch GA, Schnetter W (1983) *Bacillus*
    thuringiensis var. *tenebrionis*, a new pathotype efffective against
    larvae of Coleoptera. Z Angew Entomol 96:500-508
Krieg A, Huger AM, Langenbruch GA, Schnetter W (1984) New results on
    Bacillus thuringiensis var. *tenebrionis* with special regard to its
    effect on Colorado potato beetle. Anz Schaedlingskd Pflanzenschutz
    Umweltschutz 57:145-150
Kronstad JW, Whiteley HR (1986) Three classes of homologous *Bacillus*
    thuringiensis crystal protein genes. Gene 43:29-40
Kronstad JW, Schnepf HE, Whiteley HR (1983) Diversity of locations for
    Bacillus thuringiensis crystal protein genes. J Bacteriol 154:419-428
Luthy P, Ebersold HR (1981) *Bacillus thuringiensis* delta endotoxin:
    Histopathology and molecular mode of action, pp. 235-267. In Davidson
    EW (ed) Pathogenesis of invertebrate microbial diseases. Allanheld
    Toronto
Murphy DW, Sohi SS, Fast PG (1976) *Bacillus thuringiensis* enzyme digested
    delta endotoxin: Effect on cultured insect cells. Science
    194:954-956
Nagamatsu Y, Itai Y, Hatanaka C, Funatsu G, Hayashi K (1984) A toxic
    fragment from the entomocidal crystal protein of *Bacillus thurin-*
    giensis. Agric Biol Chem 48:611-619
Obukowicz MG, Perlak FJ, Kusano-Kretzmer K, Mayer EJ, Watrud LS (1986a)
    Integration of the delta endotoxin gene of *Bacillus thuringiensis*
    into the chromosome of root colonizing strains of pseudomonads using
    Tn5. Gene 45:327-331

Obukowicz MG, Perlak FJ, Kusano-Kretzmer K, Mayer EJ, Bolten S, Watrud LS (1986b) Tn5 mediated integration of the delta endotoxin gene from *Bacillus thuringiensis* into the chromosome of root colonizing pseudo-monads. J Bacteriol 168:982-989

Schnepf HE, Wong HC, Whiteley HR (1985) The amino acid sequence of a crystal protein from *Bacillus thuringiensis* deduced from the DNA base sequence. J Biol Chem 260:6264-6272

Schnepf HE, Whiteley HR (1981) Cloning and expression of the *Bacillus thuringiensis* crystal protein gene in *E. coli*. Proc Nat Acad Sci USA 78:2893-2897

Schnepf HE, Whiteley HR (1985) Delineation of toxin encoding segment of a *Bacillus thuringiensis* crystal protein gene. J Biol Chem 260:6273-6280

Sekar V (1986) Biochemical and immunological characterization of the cloned crystal toxin of *Bacillus thuringiensis* var. *israelensis* . Biochem Biophys Res Comm 137:748-751

Sekar V, Carlton BC (1985) Molecular cloning of the delta endotoxin gene of *Bacillus thuringiensis* var. *israelensis*. Gene 33:151-158

Shah DM, Horsch RB, Klee HJ, Kishore GM, Winter JA, Tumer NE, Hironaka CM, Sanders PR, Gasser CS, Aykent S, Siegel NR, Rogers SG, Fraley RT (1986) Engineering herbicide tolerance in transgenic plants. Science 233:478-481

Shibano Y, Yamagata A, Nakamura N, Iizuka T, Sugisaki H, Takanami M (1985) Nucleotide sequence coding for the insecticidal fragment of the *Bacillus thuringiensis* crystal protein. Gene 34:243-251

Shivakumar AG, Gundling GJ, Benson TA, Casuto D, Miller MF, Spear BB (1986) Vegetative expression of the delta endotoxin genes of *Bacillus thuringiensis* subsp. *kurstaki* in B. subtilis. J Bacteriol 166:194-204

Thorne L, Garduno F, Thompson T, Decker D, Zounes M, Wild M, Walfield, AM, Pollock TJ (1986) Structural similarity between the Lepidoptera and Diptera specific insecticidal endotoxin genes of *Bacillus thurin-giensis* subsp. *kurstaki* and *israelensis*. J Bacteriol 166:801-811

Visser B, van Workum M, Dullemans A, Waalwijk C (1986) The mosquitocidal activity of *Bacillus thuringiensis* var. *israelensis* is associated with the Mr 230,000 and 130,000 crystal proteins. FEMS Microbiol Lett 30:211-214

Waaljik C, Dullemans A, van Workum MES, Visser B (1985) Molecular cloning and the nucleotide sequence of the Mr 28,000 crystal protein gene of *Bacillus thuringiensis* subsp. *israelensis*. Nucleic Acids Res 13:8207-8217

Wabiko H, Raymond KC, Bulla LA (1986) *Bacillus thuringiensis* entomocidal protoxin gene sequence and gene product analysis. DNA 5:305-314

Ward ES, Ellar DJ (1986) *Bacillus thuringiensis* var. *israelensis* delta endotoxin. J Mol Biol 191:1-11

Ward ES, Ridley AR, Ellar DJ, Todd JA (1986) *Bacillus thuringiensis* var. *israelensis* delta endotoxin. J Mol Biol 191:13-22

Watrud LS, Perlak FJ, Tran M-T, Kusano K, Mayer EJ, Miller-Wideman MA, Obukowicz MG, Nelson DR, Kreitinger JP, Kaufman RJ (1985) Cloning of the *Bacillus thuringiensis* subsp. *kurstaki* delta endotoxin gene into Pseudomonas fluorescens: Molecular biology and ecology of an engi-neered microbial pesticide, pp. 40-46. In Halvorson HO, Pramer D, Rogul M (eds) Engineered organisms in the environment. American Society for Microbiology Washington

Wilcox DR, Shivakumar AG, Melin BE, Miller MF, Benson TA, Schopp CW, Casuto D, Gundling GJ, Bolling TJ, Spear BB, Fox JL (1986) Genetic engineering of bioinsecticides, pp. 395-413. In Inouye M, Sarma R (eds) Protein engineering. Academic Press New York

PESTICIDE USE EXPOSURE AND REGULATION IN DEVELOPED AND DEVELOPING COUNTRIES

Dr A.F. Pelfrène *
Director, Product Toxicology and Safety
Rhône-Poulenc Agrochimie
14-20, rue Pierre Baizet
69263 Lyon Cédex 09 - France

It is reasonably questionable that pesticides have brought to the world population over the last half century or so a large amount of what is usually designated as "benefits" whether from the viewpoint of general health and welfare or from the viewpoint on increased food supply of nutritional quality. However, it is also true and unquestionable that like in any other human activity, these very positive results have been accompanied by drawbacks.

Unfortunately, as we all know "good news don't sell" and therefore only the "not so positive side" of the pesticide story makes the headlines and is receiving a lot more emphasis than it really deserves based on the "hazards vs benefits" ratio of pesticide uses, and as a consequence the world has now a totally unbalanced perception of the pesticide situation. Nevertheless, this situation has triggered over the last decade or so the emergence of an international and collective awarness of the potential risks of an inconsiderate use of pesticides, while keeping in mind the beneficial aspects of their judicious uses.

Dr G. Vettorazzi **
World Health Organization
Avenue Appia
1211 Geneva 27, Switzerland

* Formerly Medical Toxicologist, Unit of Pesticide Development and Safe Use of the World Health Organization, Geneva, Switzerland

** Also Professor of Experimental Toxicology, University of MIlano, Italy

The need for a better control of the risks has become evident, initially in the industrialized countries, and later on in the developing countries as well. It was soon found that, despite all the good will and effort, this was not a simple matter because the situations in the developed and the developing countries are so different on many accounts. In order to put the problem into perspective, it may be worth giving here few examples that are often missing or ignored in the anti-pesticides diatribes. For example, the starvation and food supply issue is becoming crucial in today's world, hundred of million of people being unadequately fed while hundred of thousands or probably even millions are literally dying from starvation. This aspect is well publicized. What, on the other hand is not, is the fact that, for example in South East Asia, the population between now to the year 2000, will increase by 45 % while in the industrialized world, the growth will be only 10 %. Consequently if the food supply does not grow at an equivalent pace, dramatic starvation will appear in this region of the world. What is even more alarming is that in this region, the average surface of agricultural land per farmer is declining : in 1970 a south east asian farmer cultivated an average area of 1 ha, it is estimated that in 1982 this area was only 0.96 ha, a decline of 4 %. In Africa, the figures are respectively 1.61 ha in 1970 and 1.48 ha in 1982 or a decrease of 8 %. There is only one solution to avoid the occurence of situations of dimensions many times more dramatic than the one we presently know : to significantly increase the crop yields. An even this might not be enough since it has been estimated by the Food and Agriculture Organization of the United Nations (FAO) that, in the developing countries, the annual percentage of crop losses due to weeds, insects, plant diseases, predators such as birds and rodents was approximately 40 % meaning that the populations in those countries get only slightly more that half what they expect to grow. In order to compare, crop losses in the developed countries represent only 25 %. The situations just described here and the corresponding figures are public information, available to anyone concerned by these questions since they are published in the FAO World Food Report of 1984, and they should not be ignored.

Similarly, the situation of the most important tropical diseases again plaguing exclusively the developing countries is also dramatic (many WHO publications and documents are available on these topics) : malaria is not declining, it is at best stabilized in some countries, it is increasing in others, onchocerciasis (river blindness) control is a success in Western Africa, but this requires 22,000 km of rivers being sprayed with larvicides every weeks. Those are only two examples among many others. In the industrialized world, no such widespread endemic diseases exist.

To conclude this brief overview of the world situation, I shall say that it would appear quite obvious to a normally constituted individual that the use of pesticides - judicious use of pesticide that is - remains more than ever a necessity if we want to avoid falling inexorably into an infernal and inextricable situation in the next 25 years, or even less. This also means that, since pesticide uses are going to increase, a better control of their safety will also be required, because as it is well documented, it is known that poisoning by pesticides do occur even if the estimates of the incidence currently circulating are not really reliable. This lack of accurancy and reliability is easy to understand and this opens the door to tremendous speculations. WHO has recently published an up-dated global estimate of the number of pesticide poisonings. The Organization has cautioned, in its report of consultation on "planning strategy for the prevention of pesticide poisoning" on the interpretation of the given estimates because of the recognized difficulty in obtaining useful figures from the developing countries and even from some of the industrialized ones. To illustrate this difficulty, the report gives 4 different estimates based on different sets of data and states that the observed difference between the 1972 and the 1985 estimates probably reflects the availability of better data from some countries, the increased use of pesticides and an increase in the world's population.

Nevertheless, whatever the exact figure may be, the question of pesticide poisoning definitely deserves close attention on the part of all parties involved. And the basic framework is already in place in the industrialized countries. It is made of a strong network of regulations, guidelines, registration systems working both on a national and international (EEC, OECD) basis.

Unfortunately the situation is not as favorable in the developing world. There are reasons for that, among which the following seems to be the most obvious or the most important :
. lack of national laws and regulations and in case such laws exist, lack of enforcement.
. inexistence or inadequate institutional facilities for the registration of imported or locally manufactured active ingredients and formulations.
. general public and users unawareness of the dangers of misused pesticides.
. general lack of political will to cope with the problem.

The basic reasons outlining this situation are easy to understand. There is a chronic lack or shortage of qualified resources in those developing countries, different priorities for the allocation of scarcely available fundings and personels ; lack of education and training programes for both the regulators and the users.

There is a strong temptation at the moment on the part of both the governments of the developed countries and of pressure groups of environmentalists, antipesticides activits, etc ..., to compensate this lack of inadequancy of specific structures in many developing countries by exporting their "ready-to-use" own system, in other words to "think and decide" in their place what they should import, sell or use. I do think that this is neither appropriate nor fair. Instead those countries should be encouraged and helped designing, developing and implementing their own system, based on their own local conditions which are not, by far, identical to those of the developed countries.

It is well known that in the industrialized countries pesticides are thoroughly tested for the toxicity on mammalian systems exploring their potential efects on a large number of organs and functions ; their impact on non-target organisms is also evaluated, as well as their fate in soils, air and water. This large number of sophisticated tests is performed in highly specialized laboratories, according to stringent guidelines and protocols. The reports are carefully reviewed and evaluated by trained staff of well organized governmental bodies before any product or compound is released for public use. It takes, between the time a new molecule is synthetized by the chemist and the time it is authorized to reach the market place an average of 8 to 10 years during which period approximately 3 millions of dollars are spent on toxicology and environmental studies alone. This gives an idea of the amount of high quality resources which are required from both industry and governments to insure that the pesticides put on the market today are safe to use under the recommended conditions.

In the developing countries however as was said before, this satisfactory situation does not exist. But, apart from the administrative frame work for registering imported and locally manufactured products, there is no need to develop local generation of toxicological data which is very costly and lengthy process since those data are already available from the country of origin. What is however needed is a network of qualified and well trained toxicologists and regulator, able to evaluate the data that are submitted in order to make sound judgements on the opportunity of importing such or such compounds and to make use recommendations according to the local conditions, which may be or are often different from one country to the other. There are various reasons for such differences : economical, political, cultural, entomological, meteorological, geographical, agricultural, etc ...

This is why a general "ready-made" solution cannot be found to be applicable as such every where, most importantly a solution effective in the industrialized country cannot and must not be "exported" to a developing country where it soon becomes inefective. This is something that does not seem to be well understood by most environmentalists. This is also why, in order to bring developing countries to concentrate the scarce resources and to cut down the costs, regional (i.e., South East Asia, South America, Central American, Eastern and Western Africa, etc...) efforts should be encouraged and hel-ped. Such geographical areas usually share similar problems so similar solutions can be designed in common with the necessary help and advice of industry and governments of the industrialized countries.

In doing so, both sides (i.e., industrialized and developing world) must exercize great care not to, on the one hand, for the industrialized partners, impose unrealistic solutions onto the developing side and, on the other hand, to go for the easy solution and accept solutions that will not fit the local conditions. This is unfortunately what we are seeing taking place today. As was said earlier, the developed countries have developed a very sophisticated system for testing and controlling pesticides in their own area. And more and more developing countries are now just copying such systems. It is, in my opinion, a serious mistake. Because, first of all, these systems are very expensive, they require highly trained and skilled laboratories and personels which are lacking in the developing countries and they are based on criteria which are not those applicable in the developing countries.

For example, I would like to take a very famous scape goal, DDT. DDT has been banned years ago in many if not all developed countries. The decision to ban DDT was based at that time on the results of experimental studies showing that at high dose, DDT induced liver growths in mice. DDT was then called a carcinogen. It was also said that DDT accumulated in the body, concentrated in the food chain, etc... DDT was the best insecticide used in controlling malaria among other endemic tropical diseases which killed hundreds or thousands of people worldwide. Soon after the ban of DDT by industrialized countries, several developing countries took the same decision, based not on their own judgement or criteria, but just because such and such country did ban it ! In those developing countries, malaria was under control and the number of cases declining. Within a very short period of time after the ban of DDT, malaria was back with the number of cases raising dramatically. The reasons for this situation are quite clear : the use of DDT is not really necessary in the developed countries which are free from endemic vector borne disease.

When DDT was banned in the industrialized countries, alternative insecticides were already available, but they were, like any new compounds, much more expensive than DDT. The industrialized countries could afford the more expensive alternatives, whereas the developing countries could not and were left with nothing to control their vector borne diseases. This does not mean that the developing countries should accept, import and use any type of compounds which is proposed on the market. On the contrary, this means that they should exercize their own judgement regarding the opportunity to import or to manufacture and to use a given product, according to their own local needs and conditions. In order to do so in an effective manner, they must develop, in collaboration with the developing countries and industry which have the experience and expertise, their own system and train their own people, both pesticide regulators and users.

An excellent document, recently issued by the Food and Agriculture Organization of the United Nations (FAO) is setting the basis for such collaborative international effort. This document is the "International code of conduct on the distribution and use of pesticides". It is the fruit of thorough discussions between FAO, the WHO and other UN agencies, and NGO's (Non Governemental Organizations). This lengthy discussion has much benefited from the active participation of the international agrochemical industry. The basic objectives of this voluntary code are to set forth responsibilities and establish voluntary standards of conduct for all public and private entities engaged in or affecting the distribution and use of pesticides, and here in the most important part : "particularly where there is no or an inadequate rational law to regulate pesticides". The code describes the share responsibility of many segments of society, including governments, individually or in regional groupings, industry, trade and international institutions to work together so that the benefits to be derived from the <u>necessary</u> and <u>acceptable use</u> of pesticides are achieved without significant adverse effects on people or the environment. The code further addresses the need for a cooperative effort between governments of exporting and importing countries, industry, users, public sector organizations such environmental and consumer groups. This is more or less verbation the first three article of the code.

It is hope that all parties involved after agreeing on the basic objectives of this code will cooperate in implementing the standards set forth with fairness and objectivity for the benefit of the most needy population of the world.

SELECTED   CONTRIBUTIONS

Species Differences in Serum Paraoxonase Correlate with Sensitivity to Paraoxon Toxicity

L. G. Costa, R. J. Richter, S. D. Murphy, G. S. Omenn, A. G. Motulsky, and C. E. Furlong
Depts. of Environmental Health and Medicine (Medical Genetics) and Center for Inherited Diseases, University of Washington, Seattle, Washington (USA)

Paraoxon (diethyl-p-nitrophenylphosphate) is the active metabolite of parathion (diethyl-p-nitrophenylphosphorothioate), a widely used organo-phosphorus insecticide. After activation of parathion by the cytochrome P450 system, the paraoxon is hydrolyzed by a serum enzyme, paraoxonase (EC 3.1.1.2), to generate p-nitrophenol and diethylphosphoric acid (Zech and Zurcher, 1974). Measurement of paraoxonase activity in sera from human subjects of European origin (i.e. Caucasian) has revealed a bimodal or trimodal distribution (Mueller et al. 1983; Eckerson et al. 1983). On the basis of enzymatic tests, humans can be divided into three serum paraoxonase phenotypes: A (low activity), AB (intermediate activity), and B (high activity), with about 50% of the studied populations of European origin displaying low activity (Eckerson et al. 1983).
There is a 10-15 folds difference in serum paraoxonase activity between the extremes of the low and high activity groups. If serum paraoxonase is important in the detoxification of paraoxon, then individuals with low activity would be expected to have a diminished ability to metabolize paraoxon and, therefore, to be more sensitive to paraoxon toxicity (Omenn, 1982; La Du and Eckerson, 1984). Before epidemiological studies verifying this hypothesis are conducted, it is useful to determine in animal models whether differential levels of serum paraoxonase lead to a differential sensitivity to paraoxon toxicity. It is known that birds, which have extremely low levels of serum paraoxonase, are more susceptible to the toxic effect of paraoxon than mammals (Brealey et al. 1980). The aim of the present study was to determine whether the rabbit and the rat, two mammalian species whose serum paraoxonase activity differs by several-fold (Aldridge, 1953; Zech and Zurcher, 1974), are differently affected by paraoxon because of their difference in serum paraoxonase activity.

NATO ASI Series, Vol. H13
Toxicology of Pesticides: Experimental, Clinical
and Regulatory Aspects. Edited by L. G. Costa et al.
© Springer-Verlag Berlin Heidelberg 1987

Materials and Methods

Male Sprague-Dawley derived rats (Tyler Laboratories Inc., Bellevue, WA) and male New Zealand white rabbits (R&R Rabbitry, Stanwood, WA) were used in this study. Paraoxon (diethyl-p-nitrophenylphosphate; 95% pure) was purchased from Chem Service (West Chester, PA). The organophosphate was dissolved in corn oil and administered by i.p. injection in a volume of 1 ml/kg. Control animals were injected with corn oil only.

Serum paraoxonase activity was measured by a modification of the method of Ortigoza-Ferado et al. (1984) at pH 8.0 in the presence of 2 M NaCl and 1mM Ca++ (Furlong et al., unpublished). Cholinesterase activity was assayed by a modification of the method of Ellman (1961) as described by Costa et al. (1986).

Hydrolysis of phosphoinositides induced by the cholinergic agonist carbachol was taken as a functional index of muscarinic receptor activation. Accumulation of inositol phosphates in slices from cerebral cortex was determined according to the method of Berridge (1982) as described in detail by Costa et al. (1986).

Results and Discussion

Previous studies had reported that the rabbit had a 65-fold (Zech and Zurcher, 1974) or a 20-fold (Aldridge, 1953) higher level of serum paraoxonase than the rat. When we measured serum paraoxonase activity with the new method developed by Furlong et al. which utilizes a lower pH, and higher salt, thus eliminating the interference by albumin, (Ortigoza-Ferado et al. 1984), we found a 7-fold difference in activity between rats and rabbits. Serum paraoxonase activity (μmol PO/min/L) was 140 ± 8 in the rat and 1005 ± 48 in the rabbit (mean $\pm$ SEM; n=16).

Since high doses of paraoxon might be expected to overcome any difference in serum paraoxonase activity, we decided against the use of LD_{50} measurements as indexes of toxicity. Instead, we initiated a series of experiments aimed at determining the lowest dose of paraoxon causing some sign of cholinergic intoxication. Following numerous trials, we determined that the minimal effective doses of paraoxon were 0.5 mg/kg in the rat and 2.0 mg/kg in the rabbit. These doses have been determined after a careful titration of paraoxon, due to the steep dose-response curve of organophosphates in general, and paraoxon in particular. For example, in the rabbit a dose of 2.4 mg/kg would kill all the animals, while a dose lower than 2.0

mg/kg would not cause, or cause only occasionally, any detectable sign of intoxication. Animals were observed for six hours following injection of paraoxon. Signs of intoxication caused by 0.5 mg/kg paraoxon in the 8 rats tested were: diarrhea (8/8), chewing (7/8), abdominal twitching (7/8), shivering (6/8), yawning (4/8), tremors (2/8), urination (1/8), stretching (1/8). A dose of 2.0 mg/kg paraoxon in the rabbits caused the following signs (n=8): urination (7/8), abdominal twitching (6/8), shivering (5/8), diarrhea (1/8), sneezing (1/8).

To confirm that these two doses of paraoxon were causing similar biological effects, we measured cholinesterase activity in various tissues of control and paraoxon-treated rats and rabbits. Doses of paraoxon of 0.5 mg/kg in the rat and 2.0 mg/kg in the rabbit caused similar degree of cholinesterase inhibition. Percent inhibition was 66 and 79 (diaphragm), 83 and 96 (plasma), 60 and 76 (hippocampus) in rats and rabbits, respectively (n=3 for each group). It is striking to note that decreases of cholinesterase activity of such degree were accompanied by only slight signs of cholinergic intoxication.

Since cholinesterase activity of different species has been shown to have different sensitivity to organophosphate inhibition (Wang and Murphy, 1982), we also determined the potency of paraoxon in inhibiting rat and rabbit brain cholinesterase in vitro. The values of IC_{50} (concentration of paraoxon necessary to inhibit 50% of enzyme activity) were 5.1 ± 1.1 nM in the rat and 6.6 ± 1.4 nM in the rabbit (mean $\pm$ SEM of three determinations, each done using six concentrations of paraoxon), indicating that cholinesterase from both species had the same sensitivity to paraoxon.

While cholinesterase is the primary target for organophosphate toxicity, the signs of poisoning are due to the action of the accumulated acetylcholine on cholinergic receptors, which could be considered a secondary target for toxicity. Therefore, we examined muscarinic receptor function in rat and rabbit brain, by measuring the accumulation of inositol phosphates in cerebral cortex slices following stimulation by the cholinergic agonist carbachol. This response is thought to be coupled in cerebral cortex with the M_1 subtype of muscarinic receptor (Costa et al. 1986). There was no difference in the ability of carbachol to stimulate phosphoinositide hydrolysis in the two species. At a concentration of 10 mm, carbachol induced a 530 ± 35% increase in the accumulation of inositol phosphates in the rat and a 490 ± 50% increase in the rabbit (mean $\pm$ SEM; n=3).

These findings indicate that a 7-fold difference in serum paraoxonase results in a 4-fold difference in sensitivity to paraoxon toxicity. This difference in sensitivity is not due to the intrinsic characteristics of the primary target (cholinesterase) or a secondary target (cholinergic receptors). Moreover, as indicated by the results of Aldridge (1953), it appears that paraoxonase levels are similar between rat and rabbit in liver, kidney, spleen, and brain, thus leaving serum paraoxonase the only significant difference between the two species. In addition, another possible detoxication pathway for organophosphates (i.e. oxidative dealkylation) does not appear to be relevant in the case of paraoxon (Murphy, 1980). Furthermore, Butler et al. (1985) have recently shown that paraoxonase has a major role in the rate of disappearance of paraoxon in the rabbit. All these findings suggest that susceptibility of people to paraoxon poisoning may vary according to their inherited level and type of serum paraoxonase. Therefore, surveys of farm workers for paraoxonase levels and symptoms of organophosphate toxicity are highly indicated.

Acknowledgments

This study was supported in part by grants from the Charles A. Dana Foundation and the National Institutes of Health (ES-03424, GM-15253, and OH-00054). Ms. Grace Kaylor provided valuable technical assistance.

References

Aldridge WN. Biochem. J. 53:117-124, 1953.
Berridge MJ, Downes CP, Hanley MR. Biochem. J. 206:587-595, 1982.
Brealey CJ, Walker CH, Baldwin BC. Pestic. Sci. 11,546-554, 1980.
Butler EG, Eckerson HW, La Du BN. Drug Met. Dispos. 13:640-645, 1985.
Costa LG, Kaylor G, Murphy SD. J. Pharmacol. Exp. Ther. 239:32-37, 1986.
Eckerson HW, Wyte CM, La Du BN. Am. J. Hum. Genet. 35:1126-1138, 1983.
Ellman GL, Courtney KD, Andres V, Featherstone RM. Biochem. Pharmacol. 7:88-95, 1961.
La Du BN, Eckerson HW. IN Genetic Variability in Responses to Chemical Exposure (GS Omenn, HV Gelboin, eds.) Cold Spring Harbor Laboratory, New York, 1984, pp. 167-175.
Murphy SD. IN Casarett and Doull's Toxicology: The Basic Science of Poisons (J Doull, CD Klaassen, MO Amdur, eds.) MacMillan, NY, 1980, pp. 357-408.
Omenn GS, J. Occup. Med. 24:369, 1982.
Ortigoza-Ferado J, Richter RJ, Hornung SK, Motulsky AG, Furlong CE. Am. J. Hum. Genet., 36:295-305, 1984.
Wang C, Murphy SD. Toxicol. Appl. Pharmacol. 66:409-419, 1982.
Zech R, Zurcher K. Comp. Biochem. Physiol. 48B:427-433, 1974.

The Effects of Pyrethroid Insecticides on Synaptic Transmission in Slices of Guinea Pig Olfactory Cortex

J.T. Eells, S. Watabe,[1] N. Ogata,[1] and T. Narahashi[1]
Department of Pharmacology
Medical College of Wisconsin
8701 Watertown Plank Road
Milwaukee, Wisconsin 53226 (USA)

INTRODUCTION

The synthetic pyrethroids are potent and widely used insecticides. Pyrethroids modify the ionic permeability of nerve membranes and produce distinctive poisoning syndromes in insects and mammals (1,4,5). These agents have been divided into two major classes on the basis of their structures, neurophysiological and toxicological actions (2). Pyrethroids containing an α-cyano substituent have been classified as type II pyrethroids and include the agents deltamethrin and fenvalerate. Type II pyrethroid insecticidal action is associated with long nerve discharges and nerve membrane depolarization and these compounds produce a poisoning syndrome in mammals characterized by tremors, choreoathetosis and tonic seizures. The noncyano pyrethroids (type I) include the agents permethrin and tetramethrin and produce extensive repetitive firing in isolated invetebrate nerve preparations. The type I poisoning syndrome in mammals is characterized by hyperactivity and tremor.

The actions of pyrethroids have been well characterized in the invertebrate nervous system, however, mammalian studies have thus far provided little fundamental information on pyrethroid action (4). The _in vitro_ brain slice is an ideal preparation in which to study the actions of pyrethroids at a level of organization between the isolated nerve membrane and the intact animal. The objective of this investigation was to characterize the actions of type I and type II pyrethroids on synaptic activity in the guinea pig olfactory cortex.

METHODS

Experiments were performed on superficial slices (400–600 μm) of

[1] Department of Pharmacology, Northwestern University Medical School, 303 East Chicago Avenue, Chicago, Illinois 60611 (USA)

NATO ASI Series, Vol. H13
Toxicology of Pesticides: Experimental, Clinical and Regulatory Aspects. Edited by L. G. Costa et al.
© Springer-Verlag Berlin Heidelberg 1987

guinea pig olfactory cortex. Slices were continuously perfused at a rate of 2 ml/min with oxygenated incubation medium at 30–32°C. The composition of the incubation medium was: (mM) NaCl, 124; NaHCO$_3$, 13; KCl, 5; CaCl$_2$, 2.6; KH$_2$PO$_4$, 1.24; MgSO$_4$, 1.3; and glucose, 10. The lateral olfactory tract (LOT) was stimulated through a bipolar electrode composed of twisted silver wire (50 microns in diameter) insulated except at the tip. Single or paired electrical pulses 100 μsec in duration and 0.5–20 V in intensity were delivered at a frequency of 0.4 Hz. Responses were recorded from the prepiriform cortex using glass microelectrodes filled with 0.9% NaCl (D.C. resistance 1–3 megohm). An indifferent electrode of Ag/AgCl was in the bottom of the chamber. All measurements of the N-wave of the evoked potential were made from the recording baseline to the peak of negativity. Deltamethrin was obtained from Rousel Uclaf (Marseille, France). Fenvalerate and tetramethrin isomers were provided by the Sumimoto Chemical Company (Takarazuka, Japan). Stock solutions of pyrethroids were prepared in dimethylsulfoxide (DMSO) (Sigma Chemical Co., St. Louis, MO) and added to the perfusion medium at fixed concentrations (0.1–10 μM) for 10 minutes.

RESULTS

The evoked potential (EP) recorded from the prepiriform cortex in response to stimulation of the LOT with paired electrical pulses (8 V, 50 msec interval) is shown in Figure 1 (top panel). This potential is comprised of a negative wave (N-wave) with an amplitude of 1–3 mV and latency of 4–6 msec (to peak) upon which is superimposed 1 or 2 positive peaks. The negative wave is a reflection of the population excitatory postsynaptic potential (EPSP) arising from the superficial synapses that are activated by the LOT fibers (3,6). The positive peaks are population spikes and reflect the

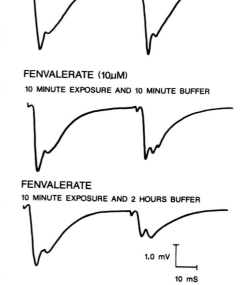

FENVALERATE (10μM)
10 MINUTE EXPOSURE AND 10 MINUTE BUFFER

FENVALERATE
10 MINUTE EXPOSURE AND 2 HOURS BUFFER

1.0 mV

10 mS

Fig 1: *Effect of Fenvalerate on the Evoked Potential.*

synchronous discharge of monosynaptically activated cortical neurons (3,6).

Figure 1 (middle panel) shows the response recorded from the same slice 20 min after exposure to fenvalerate (10 μM for 10 min) and (bottom panel) shows the response 2 hours after fenvalerate exposure. Fenvalerate markedly attenuated the amplitude of the second (test) evoked potential in paired pulse protocol reducing the EP ratio (test EP/conditioning EP) from 1.11 ± 0.01 (SEM) (n=6) in the control to 0.81 ± 0.04 (n=6), 20 minutes post fenvalerate and to 0.40 ± 0.03 (n=4) two hours after pyrethroid exposure.

Type I and type II pyrethroids had similar effects on the evoked potential in the guinea pig olfactory cortex. Figure 2 shows the effect of tetramethrin (type I), fenvalerate and deltamethrin (type II) on the amplitude of the first (conditioning) potential (A) and on the EP ratio (B). Both classes of pyrethroids produced a slight reduction in the amplitude of the conditioning evoked potential and a marked reduction in the amplitude of the test (second) evoked potential. Pyrethroid action was time-dependent, concentration-dependent and not reversed by extensive washing in pyrethroid-free medium.

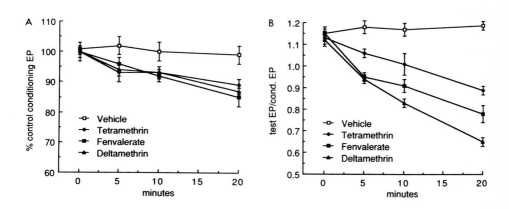

Fig 2: Effect of Type I and Type II Pyrethroids on the Conditioning Evoked Potential and on the EP Ratio

The effect of deltamethrin on the amplitude of the N-wave at different stimulus intensities is shown in Figure 3. Stimulus intensity was varied from 1-20 V and the cortical response was expressed as percent of the maximal response in the control preparation. A significant reduction in the EP at low stimulus intensities was observed following deltamethrin exposure (10

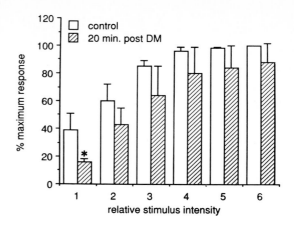

Fig 3: *Effect of Deltamethrin on the Evoked Potential at Different Stimulus Intensities*

µM). As stimulus intensity was increased the response in the pyrethroid-treated preparations increased to the same maximal response seen in the control preparations. Further studies are necessary to define this shift in sensitivity, however, these data do suggest that deltamethrin increases the threshold of stimulation required to elicit the EPSP.

Pyrethroid action on the evoked potential following paired pulse stimulation was more pronounced at short interpulse intervals as shown in Figure 4. In these experiments the interpulse interval was varied from 20 - 100 msec at a supramaximal stimulus intensity. Under control conditions paired pulse facilitation was observed at all interpulse intervals of 40 msec or greater with an EP ratio between 1.1 - 1.25. The EP ratio after exposure to deltamethrin (DM) did not

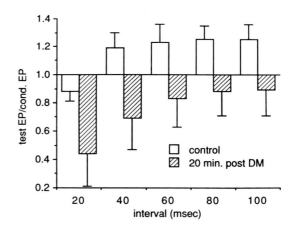

Fig 4: *Effect of Deltamethrin on the Evoked Potential at Different Interpulse Intervals*

exceed 1.0 and was profoundly depressed at the 20 and 40 msec intervals.

271

DISCUSSION

The progression of events from pyrethroid induced alterations in neuronal ionic fluxes to the poisoning syndrome observed in the intact animal is not understood. We have examined the actions of type I and type II pyrethroids on synaptic function in slices of guinea pig olfactory cortex. When the LOT was stimulated at a frequency of 0.4 Hz (2.5 sec separation), deltamethrin increased the stimulus threshold required to elicit a maximal response. Both classes of pyrethroids depressed the second evoked potential when paired-pulse stimulation was employed. In the case of deltamethrin, this paired pulse-inhibition was also frequency-dependent. The pyrethroid-induced depression of the evoked potential may reflect a reduction in synaptic input from the LOT to the cortical neurons. In non-mammalian nerve preparations, pyrethroids have been shown to depolarize the nerve membrane resulting in an inhibition of evoked activity and a frequency-dependent conduction block (1,4,5). In repetitively stimulated frog myelinated nerve fibers, Vijverberg and van den Berken (5) have shown that type II pyrethroids produce a summation of after potentials leading to depolarization of the nerve membrane and to a depression of the action potential amplitude. The degree of depression depended on the stimulation frequency and the action potential regained full amplitude if the nerve was left unstimulated for a few seconds. The similarity between these findings and ours lends support to the hypothesis that pyrethroids alter sodium conductance and depolarize the presynaptic or axonic membrane.

REFERENCES

1. Narahashi, R. and Lund, A.E. (1980) Insect neurobiology and pesticide action. Society of the Chemical Industry, London.
2. Narahashi, T. (1985) NeuroToxicology 6: 3-32.
3. Richards, C.D. and Sercombe, R. (1968) J. Physiol. 197: 667-683.
4. Vijverberg, H.P.M. and van den Berken, J. (1982) Neuropathol. and Appl. Neurobiol. 8: 421-440.
5. Vijverberg, H.P.M. and de Weille, J.R. (1985) NeuroToxicology 6: 23-34.
6. Yammamoto, C. and McIlwain, H. (1966) J. Neurochem. 13: 1333-1343.

Critical review of Pesticide-related Alimentary Outbreaks.

A. FERRER
SERVICIO DE TOXICOLOGIA
HOSPITAL CLINICO UNIVERSITARIO
ZARAGOZA. ESPAÑA.

CABRAL, J.R.P.

IARC; 150 Cours Albert-Thomas
69372 LYON. FRANCE.

INTRODUCTION

It is possible to distinguish two types of collective intoxications by pesticides or thet by-products: Epidemic situations and catastrophic situations.

The term epidemic has been taken from the infectious field to qualify a pathological outbreak affecting a significant number of population in a limited period of time with a common toxic source.

The first toxic alimentary epidemics reported in scientific literature were due to a vegetal source: the rye ergot attacking important areas of Europe in the 19th century.

During the 20th century the chemical and particularly the synthetic pesticides has been the main causes of those accidents.

In this field you can find two kinds of references dealing with accidental and occupational problems in the literature (HAYES 1975 and 1982).

The catastrophic situations are, in general, of industrial origin, their main characteristics being the suddenness and severity of a toxic escape to the environment with human health and ecological implications. In recent years the attention of the media has been attracted because of the events in Bohpal and Seveso (GRAZIA, 1985; HOMBERBERG et al., 1979).

The usefulness of chemical pesticides is beyond doubt in the fight against crops damage and disease vectors. In spite of that, it seems necessary to maintain a constant vigilance on their acute and chronic toxicity, their carcinogenic and mutagenic potentiality and their environmental impact in order to obtain a correct and accurate risk-benefit balance.

NATO ASI Series, Vol. H13
Toxicology of Pesticides: Experimental, Clinical
and Regulatory Aspects. Edited by L. G. Costa et al.
© Springer-Verlag Berlin Heidelberg 1987

MATERIAL

From this point of view we have studied 22 pesticide-related alimentary epidemics. In them we have considered the following factors:

Toxic agent

Contamination and intoxication sources

Epidemic period

Latency period

Number of intoxicated patients

Number of deaths

Primary and major symptoms

Primary clinical diagnosis

Treatment

Sequels and congenital implications

Type, speed and efficiency of the adopted measures.

RESULTS AND DISCUSSION

On the basis of the toxic source we have classified those outbreaks into 3 groups:

1° Intoxications from foodstuffs contaminated in transportation or storage.

2° Intoxications from the consumption of seed treated for sowing.

3° Intoxications from a mistaken substance for the toxic agent when added to the food.

Each one of those groups presents significant characteristics in homogeneity which implies some specific preventive approachs.

First group

Here we have included 10 accidents having taken place in the following years and countries:

1956	Great Britain	(DAVIS and LEWIS, 1956)
1958	Indian	(KARUNAKARAN, 1958)
1958	Egypt	(WISHASI et al, 1958)
1959	Singapoore	(KANAGARATNAM et al, 1960)
1966	Egypt	(COBLE et al, 1967)
1967	Quatar	(WEEKS, 1967)
1967	Saudi Arabia	(WEEKS, 1967)

1967   Mexico   (MARQUEZ et al, 1968)

1968   USA   (OLDER and HATCHER, 1969)

1976   Jamaica   (DIGGORY et al, 1977)

Their main characteristics are:

TOXIC AGENT	Endrin	4
	Parathion	5
	Carbophenothion	1
CONTAMINATED FOODSTUFF	Flour	8
	Barley	1
	Wheat	1
CONSUMED PRODUCTS	Bread	6
	Others	4
CONTAMINATION SPOT	Ship	5
	Others	5

EPIDEMIC PERIOD	1 day–1 month
LATENCY PERIOD	24 hours
INTOXICATED PATIENTS	2.294
NUMBER OF DEATHS	178
PRIMARY DIAGNOSIS	Alimentary intoxication
PRIMARY SYMPTOMS	Nausea, vomiting
ANALYTIC DIAGNOSIS	+
TREATMENT	Symtomatic, Atropine, PAM
SEQUELS AND CONGENITAL	
IMPLICATIONS	non reported

By this means, the common patterns of the group are the nature of the toxic agent – some of the more widely used insecticides – , the contaminated foodstuffs – cereals – , the relatively short length of the epidemic and latency periods and also, the contamination spot – mainly ships–

Second group

The 7 outbreaks have happened in the following years and countries:

1932   USA (MUNCH et al, 1933)

1955/59 Turkey (SCHMID, 1960; CAM, 1960; CRIPPS et al, 1980)

1955/56   Irak   (JALILI and ABBASI, 1961)

1959/60   "   "

1961   Pakistan   (HAQ, 1963)

1965   Guatemala   (ORDOÑEZ et al 1966)

1971/72   Irak   (BAKIR et al, 1973)

Their main characteristics are:

TOXIC AGENT	Organic Hg	5
	HCB	1
	Thallium sulphate	1
CONTAMINATED PRODUCT	Seed	7
CONSUMED PRODUCT	bread	4
	others	3
EPIDEMIC PERIOD	12 days-5 years	
LATENCY PERIOD	15 days-6 months	
INTOXICATED PATIENTS	10.000	
NUMBER OF DEATHS	830	
PRIMARY DIAGNOSIS	Wrong in most cases	
PRIMARY SYMPTOMS	unspecific	
ANALYTIC DIAGNOSIS	+	
TREATMENT	Symptomatic, BAL.	
SEQUELS	Central and Peripheral nervous system affected : Hg	
	Severe mutilating scars : HCB	
CONGENITAL IMPLICATIONS	Severe brain damage : Hg	

Third group

We have included 5 cases having appeared in the following years and places:

```
1942    USA     (LIDBECK et al, 1943)
        Iran    (MORTON, 1945)
1952    Taiwan    (HSIEH, 1954)
1963    Israel    (LEWI and BAR-KHAYIM, 1964)
1977    Jamaica    (LIDDLE et al, 1979)
```

Their main characteristics are:

TOXIC AGENT	Inorganic pesticides	3
	DDT	1
	Carbamates	1
CONSUMED PRODUCTS	Divers	
CONTAMINATION SOURCE	Confusion with alimentary powder	
EPIDEMIC PERIOD	1-2 days	
LATENCY PERIOD	6 hours	
INTOXICATED PATIENTS	469	
NUMBER OF DEATHS	69	
PRIMARY DIAGNOSIS	Toxic gastroenteritis	

PRIMARY SYMTOMS	Nausea, vomiting
ANALYTICAL DIAGNOSIS	+
SEQUELS AND CONGENITAL	
IMPLICATIONS	Non reported

The specific measures to be adopted in each group are:

Shipment and storage surveillance in the first group.

Strict control of the treated seed in the second group.

and Modification of the organoleptic characteristics in the third group.

Apart from all those measures the most important point to remember is the possibility of the appearance in order to maintain an alert dispositive able to detect and control them as soon as possible.

From another point of view it would be important to bear in mind that this kind of epidemic is an excellent though undesired way for the study of the short and long term effects of pesticides in men which would be extremely beneficial if this type of structure were available.

BIBLIOGRAPHIE

Bakir, F. et al (1973) Methylmercury poisoning in Iraq. Science, 181:230-240

Cam, C. (1960) Une nouvelle dermatose epidemique des enfants. Ann Der 87/4: 393-7

Coble , Y. et al. (1967) Acute endrin poisoning. JAMA 202: 489-93

Cripps, D.J. et al. (1980) Porphiria turcica. Twenty years after HCB intoxication. Arch Dermatol, 116: 46-50

Davis, G.M. and Lewis, I. (1956) Outbreak of food-poisoning from bread made of chemical contaminated flour. Br Med J, 2: 393-8

Diggory, H.J. et al. (1977) Fatal Parathion poisoning caused by contamination of flour in international comerce. Am J Epidemiol, 106/2:145/53

Grazia, A. (1985) A cloud over Bohpal. Kalos Foundation. CIPA. New York.

HAQ, J.V. (1963) Agrosan poisoning in man. Br Med J, 1: 1579-82

Hayes, W.J. (1975) Toxicology of pesticides. W and W. Baltimore.

Hayes, W.J. (1982) Pesticides studies in man. W and W. Baltimore.

Homberger, E. et al. (1979) The Seveso accident:its nature, extent and consequences. Ann Occup Hyg, 22: 327-70

Hsieh, H.C. (1954) DDT intoxication in a family in Southern Taiwan. Arch Ind Hyg, 10: 334-46

Jalili, M.A. and Abbasi, A.H. (1961) Poisoning by ethyl mercury toluene Sulphonanilide. Br J Ind Med, 18: 303-8

Kanagaratnam, K. et al. (1960) Parathion poisoning from contaminated barley. Lancet, 1: 538-42.

Karunakaran, C.O. (1958) The Kerala food poisoning. J Indian Med Assoc 31: 204-7

Lewi, Z. y BAR-KHAYIM, Y (1964) Food-poisoning from barium carbonate. Lancet, 2: 342-3

Lidheck , W.L. et al. (1943) Acute sodium fluoride poisoning. JAMA, 121: 826-7

Liddle, J.A. et al. (1979) A Fatal episode of accidental Methomyl Poisoning Clin Toxicol, 15/2: 159-67

Marquez Mayaudon, E. et al. (1968) Problemas de contaminación de alimentos con pesticidas. Caso Tijuana. Salud Publica Mex, 10: 293-300

Morton, W. (1945) Poisoning by barium carbonate. Lancet, 2: 738-9

Munch, J.C. et al. (1933) The thallotoxicosis outbreak in California. JAMA, 100: 1315-19.

Older, J.J. and Hatcher, R.L. (1969) Food poisoning caused by carbopheno-thion. JAMA, 209: 1328-30.

Ordoñez, J.V. et al. (1966) Estudio epidemiológico de una enfermedad considerada como encefalitis en la región de los altos de Guatemala. Bol of Sanit Panam, 74: 93-107.

Schmid, R. (1960) Cutaneous porphiria in Turkey. N Engl J Med, 263:397/8

Weeks, D. E. (1967) Endrin food poisoning. A report of flour outbreaks caused by two separate shipments of endrin-contaminated flour. Bull WHO, 37: 499-512.

Wishahi, A. et al. (1958) Parathion poisoning (phosphorous compound). A report on 22 children in an outbreak. Arch Pediatr, 75: 387-96.

Peripheral and Central Enzyme Inhibition by Fenthion, Fenitrothion and Desbromoleptophos

B. Magnus Francis and M. Farage-ElAwar
Institute for Environmental Studies
Department of Veterinary Biosciences
University of Illinois at Urbana-Champaign
Urbana IL 61801
U.S.A.

INTRODUCTION

Fenthion [FEN] is suspected of causing severe and long lasting nervous system damage in exposed humans and animals, but has not been shown to cause paralysis in hens at sublethal doses (3, 7). In addition, it does not inhibit chick brain neurotoxic esterase (NTE) sufficiently to be categorized as a delayed neurotoxicant on this basis (2). Since organophosphorus ester induced delayed neurotoxicity (OPIDN) is a peripheral rather than a central paralysis (8), it was important to determine whether FEN affected peripheral more than central NTE. We therefore evaluated the effects of FEN on NTE in lympho-cytes, validating this assay by including two organophosphorus compounds (OPs) of known neurotoxic potential. Desbromoleptophos (DBL), causing paralysis in hens after a single oral dose of 20-30 mg/Kg (4), was the positive control; and fenitrothion (FTR), which does not cause delayed paralysis (9), was the negative control. The toxicity of these three OP's was compared using 4 assays: inhibition of acetylcholinesterase (AChE) in brain and in red blood cells (RBCs); inhibition of NTE in brain and in lymphocytes. The OP's were administered at doses causing overt cholinergic symptoms in chicks (FEN, FTR) or paralysis in hens (DBL). Chicks were treated per os with 1 or 2 doses of 5 mg/Kg FEN; 75 mg/Kg FTR or 50 mg/Kg DBL.

METHODS

Hybrid female broiler chicks from the University of Illinois Poultry farm were raised as described elsewhere (2), and supplied ad libitum with water and a standard chick starter diet (Purina). Treated and control chicks were matched by weight before dosing, and thereafter kept in separate cages.

FTR [O,O-dimethyl O-3-methyl-4-nitrophenylphosphorothionate, purity 95%] FEN [O,O-dimethyl O-3-methyl-4-(methylthio)phenylphosphorothionate], both from Chem Service Inc, West Chester PA; and DBL [O-2,5-dichlorophenyl O-methyl phenylphosphonothionate, purity > 98%] from the laboratory of R. L. Metcalf, were dissolved in corn oil and administered per os (po) by microliter syringe to 10, 26 or 75 day old chicks, as one or two doses given 24 hours apart. Chicks were killed by decapitation 24, 48 or 72 hours after the last dose. Brains were quickly removed, freed from blood and meninges, and homogenized, but homogenates were not centrifuged (6).

To obtain lymphocytes, blood was drawn into heparinized syringes, diluted 1:1 (v/v) with Hanks balanced salt solution (HBSS), layered over Ficoll-Paque density medium, and centrifuged at 400xg for 30 minutes at room temperature. Lymphocytes were collected from the interphase, washed twice by suspension in HBSS and centrifugation at 400xg for 10 minutes. Cell pellets were resuspended in NTE buffer, and cell number and viability were determined by trypan blue exclusion. Cells were disrupted by homogenization and by 5 passages through a 26 gauge needle. The final dilution was equivalent to 2.0×10^7 lymphocytes/ml.

Red blood cells (RBCs) were obtained by centrifuging 1 ml heparinized blood at 1000xg for 10 minutes and removing the supernatant. The pellet of RBCs was washed twice with HBSS. A 1/1000 dilution of RBC in phosphate buffer (0.1M, pH 8.0) was used for AChE assays.

NTE was assayed using O-2,6-dichlorophenyl-O-methyl phenylphosphono-thionate as NTE inhibitor (10). NTE activity was calculated as nmoles phenylvalerate hydrolyzed/min/gm tissue and is expressed as percent inhibition of control activity. Negative values are reported quanti-tatively for 10 day old chicks.

AChE assays were performed using a 1% homogenate of the brain (1). AChE activity was calculated as umoles acetylthiocholine hydrolyzed/min/gm (1) tissue, and expressed as percent inhibition of control activity. Negative values were not obtained for AChE inhibition in this study.

RESULTS AND DISCUSSION

When chicks were treated with 2 doses of FEN, the 2nd dose had to be reduced or all chicks died. Nevertheless, brain NTE was not significantly inhibited by either 1 or 2 doses of FEN in 26- or 75-day old chicks. Similarly, the much less acutely toxic FTR did not inhibit NTE in chicks of either age, whether 1 or 2 doses of 75 mg/kg were given. In contrast, the known neurotoxicant DBL inhibited brain NTE by greater than 90% after 1 or 2 doses in chicks at either 26 or 75 days of age. Inhibition of lymphocyte (WBC) NTE was somewhat less than inhibition of brain NTE after DBL exposure, and insignificant after FEN and FTR exposure. These results are shown in Table 1.

COMPOUND	AGE IN DAYS	# OF DOSES	TOTAL DOSE (mg/kg)	PERCENT ENZYME INHIBITION			
				NTE		AChE	
				BRAIN	WBC	RBC	BRAIN
FENTHION:	75	1	5.0	0	0	--	54
		2	6.7	0	27	--	61
	26	1	4.5	1.9	0	67	70
		2	6.7	0	0	45	42
FENITROTHION	75	1	75	0	0	--	45
		2	152	9.0	7	--	74
	26	1	69	0	0	76	74
		2	147	0	0	55	73
DESBROMO-LEPTOPHOS	75	1	54	97	83	--	9
		2	108	90	84	--	47
	26	1	47	98	77	33	61
		2	100	97	--	69	67

[a] too few lymphocytes for assay

TABLE 1. Enzyme inhibition 24 hours after treatment of chicks with 1 or 2 doses of FEN, FTR, or DBL

When enzyme inhibition was monitored for 72 hours following single po doses of the 3 OPs, similar results were obtained (Table 2). No NTE inhibition was seen 24 hours after FTR or FEN exposure. After 48 hours, however, a rebound of NTE to greater than control levels (negative inhibition) was seen in brains of both FTR and FEN exposed chicks. A similar overcompensation occurred in lymphocytes at 48 hours after FEN exposure and 72 hours after FTR exposure. Such rebound effects have been seen in our other studies with chicks (2), and do not seem to require an initial inhibition of the NTE. Overcompensation of AChE inhibition has also been noted in AChE levels in mallard ducklings exposed to OPs (5), but was not seen in the relatively short observation period used here. DBL caused highly significant NTE inhibition in both brain and lymphocytes, with gradual recovery of both brain and peripheral NTE during the 72 hour observation period (Table 2).

The severe inhibition of both RBC and brain AChE by FEN was comparable to the AChE inhibition caused by DBL and by FTR, and demonstrates that the lack of brain NTE inhibition after FEN treatment is not due to inability of FEN to penetrate the central nervous system. We therefore conclude that FEN does <u>not</u> inhibit NTE, and that its induction of abnormal gait in chicks is due to a mechanism other than NTE inhibition.

COMPOUND	HOURS SINCE DOSE	DOSE [mg/kg]	PERCENT ENZYME INHIBITION			
			NTE		AChE	
			BRAIN	WBC	RBC	BRAIN
FENTHION	24	5	0	a	69	43
	48		-11	-16	58	45
	72		-12	0	46	5
FENITROTHION	24	75	6	a	66	61
	48		-22	- 2	63	64
	72		0	-43	57	40
DESBROMO-LEPTOPHOS	24	50	85	60	61	55
	48		82	46	51	21
	72		53	2	12	28

[a] too few lymphocytes for assay

TABLE 2. Enzyme inhibition 24-72 hours after treatment of 10 day old chicks with 1 dose of FEN, FTR, or DBL

ACKNOWLEDGEMENTS

We wish to thank Dr. R. L. Metcalf for providing desbromoleptophos for the treatments, as well as 2,6-desbromoleptophosoxon and phenylvalerate for the NTE assays; and Dr. L.G. Hansen for his valuable advice. This work was supported by the Toxicology Program of The Institute for Environmental Studies and by The Research Board of the University of Illinois.

REFERENCES

1. Ellman GL, Courtney K, Featherstone, R (1961) A new and rapid determination of acetylcholinesterase activity. Biochem Pharmacol 7:88-95
2. Farage-ElAwar M, Francis BM (1987) J Toxicol Environ Health (in press) Effects of three organophosphate insecticides on gait, acetylcholinesterase, and neurotoxic esterase in chicks.
3. Francis BM, Metcalf RL, Hansen LG (1985) Toxicity of organophosphorus esters to laying hens after oral and topical administration J Environ Sci Health B20:73-95
4. Francis, BM (1983) Effects of dosing regimens and routes of administration on organophosphorus ester induced delayed neurotoxicity. Neurotoxicol 4:139-146
5. Hoffman DJ, Eastin WC Jr (1981) Effects of malathion, diazinon, and parathion on mallard embryo development and cholinesterase activity. Environ Res 26: 472-485
6. Johnson MK (1977) Improved assay of neurotoxic esterase for screening organophosphates for delayed neurotoxicity potential. Arch Toxicol 37:113-115
7. Leifson O (1985) Notice of proposed decisions concerning reevaluation of pesticide products 9 April 1985. California Department of Food and Agriculture, Pesticide Registration Unit
8. Metcalf, R.L. 1982. Historical perspective of organophosphorus ester-induced delayed neurotoxicity. Neurotoxicol 3(4):269-284.
9. Ohkawa H, Oshita H, Miyamoto J (1980) Comparison of inhibitory activity of various organophosphorus compounds against acetylcholinesterase and neurotoxic esterase of hens with respect to delayed neurotoxicity. Biochem Pharmacol 29:2721-2727
10. Reinders JH, Hansen LG, Metcalf RL (1983) In vitro neurotoxic esterase assay using leptophosoxon analogs as inhibitors. Toxicol Lett 17:107-111

APPLICATION OF STRUCTURE-ACTIVITY ANALYSIS FOR ESTIMATION OF POTENTIAL EFFECTS OF PESTICIDES ON ENVIRONMENTAL BIOLOGICAL TARGETS

S. Marchini*, L. Passerini, D. Cesareo* and M.L. Tosato
Department of Comparative Toxicology and Ecotoxicology
ISTITUTO SUPERIORE DI SANITA' - Viale Regina Elena, 299 - 00161 ROMA, Italy

Lack of information concerning toxic effects of pesticides on environmental biological targets, which are necessary for hazard assessment, may be tentatively filled making use of structure-activity analysis; application of the latter requires that adequate data for the relevant endpoints are available for a number of structurally similar chemicals, that elicit the same kind of toxic activity as the untested ones.

The 2-X-4,6-bis(alkylamino)-1,3,5-triazines listed in Table I (X=Cl, OCH_3 or SCH_3) are examples of pesticides that are closely similar in structure and have herbicidal activity, elicited through inhibition of photosyntesis. Available physicochemical and degradation data of some chemicals in this class permit to anticipate their common likelyhood to contaminate environmental waters and to be highly persistent; therefore, data on toxic effects on aquatic organisms are important for preliminary assessment of their potential environmental hazard. In the literature, phytotoxicity data are available for tens of s-triazine derivatives (Ebert R. and Dumfort S.W., 1976); furthermore, acute effects data on Daphnia magna have been recently developed in our laboratory for 14 s-triazines (Marchini S. et al, 1986).

These two sets of data form adequate bases for structure-activity analysis. In this paper we will show two examples of how these data may be analyzed in order to develop means to estimate toxic effects of chemicals in the same class. Some results of a multivariate statistical approach for quantitative predictions of phytotoxicity in terms of molecular descriptors, and outcomes of a qualitative approach for estimations of toxicity to Daphnia magna in terms of structural features, will be described.

---

* The author acknowledges the support given by the Commission of the European Communities for partecipation in the NATO-ASI Course where this work was presented.

MATERIAL AND METHODS

The considered chemicals are listed in Table I.

Phytotoxicity data, expressed as pI_{50} (the inverse logarithm of the molar concentration causing 50% inhibition of photosyntesis), on 19 triazines were analyzed in terms of eight molecular descriptors, namely, molecular weight, melting point, water solubility, dissociation constant, a gas chromatographic and two HPLC retention indices, and the anodic half wave potential. The PLS (Partial Least Squares) method in latent variables, which is appropriate for analysis of the above multivariate data (as it is unaffected by their eventual collinearity), was the chemometric tool adopted for developing a quantitative model for predictions of pI_{50}. Sources of data and description of the PLS method are reported in Cesareo D. et al. (1986).

Acute toxicity data on Daphnia magna , expressed as 48 h EC_{50} (mg/liter of test compound producing 50% immobilization) were measured according to standard procedures (OECD Test Guideline n.202); however, no solvent was used to increase the water solubility so that only lower limit values are reported for 3 chemicals .

EC_{50} values were analysed qualitatively with reference to the substitution pattern at the heteroaromatic ring, for the purpose of identifying possible contribution to the toxic activity of each X, R_1-R_4 substituent, as such, and in combination.

RESULTS AND DISCUSSION

The basic structure of the considered triazines, the substituent groups X, R_1-R_4, that modulate their chemical properties and biological activities, and the toxicity data, pI_{50} and EC_{50} (48 h) (TABLE I) provide the essential information needed for present study, which resulted in the development of quantative structure-activity relationships (QSAR) for predictions of phythotoxic activity, and qualitative empirical criteria for possible estimates of toxicity to Daphnia, based on structural features such as the type and characteristics of substituent groups.

The two approaches are separately dealt with in the following.

QSARs for predictions of pI_{50}.

For the purpose of QSARs, the molecular structures were firstly translated into numerical descriptors, i.e. the values of the 8 above mentioned molecu-

TABLE I General formula, substituent groups, and toxicity data, pI_{50} and 48 h EC_{50} (see text), for the considered triazines

GENERAL	N	TRIVIAL NAME	X	R_1	R_2	R_3	R_4	pI_{50}	EC_{50}
FORMULA	1	Propazine	Cl	H	$CH(CH_3)_2$	H	$CH(CH_3)_2$	5.67	11
	2	Terbutylazine	Cl	H	C_2H_5	H	$C(CH_3)_3$	6.45	>5
	3	Atrazine	Cl	H	C_2H_5	H	$CH(CH_3)_2$	5.82	>39
	4	Simazine	Cl	H	C_2H_5	H	C_2H_5	5.74	>3.5
	5	Norazine	Cl	H	CH_3	H	$CH(CH_3)_2$	5.50	-
	6	Cyanazine	Cl	H	C_2H_5	H	$C(CH_3)_2CN$	6.37	35.5
	7	Trietazine	Cl	H	C_2H_5	C_2H_5	C_2H_5	>3	6
	8	Ipazine	Cl	H	C_2H_5	C_2H_5	$CH(CH_3)_2$	3.95	-
	9	Prometone	OCH_3	H	$CH(CH_3)_2$	H	$CH(CH_3)_2$	4.00	38
	10	Terbutone	OCH_3	H	C_2H_5	H	$C(CH_3)_3$	6.56	44
	11	Secbumetone	OCH_3	H	C_2H_5	H	$CH(CH_3)C_2H_5$	5.60	-
	12	Atratone	OCH_3	H	C_2H_5	H	$CH(CH_3)_2$	5.55	-
	13	Sinetone	OCH_3	H	C_2H_5	H	C_2H_5	4.00	-
	14	Prometryne	SCH_3	H	$CH(CH_3)_2$	H	$CH(CH_3)_2$	6.67	9.7
	15	Desmetryne	SCH_3	H	CH_3	H	$CH(CH_3)_2$	5.85	26
	16	Ametryne	SCH_3	H	C_2H_5	H	$CH(CH_3)_2$	6.44	40
	17	Terbutryne	SCH_3	H	C_2H_5	H	$C(CH_3)_3$	7.15	7.1
	18	Simetryne	SCH_3	H	C_2H_5	H	C_2H_5	6.52	-
	19	Aziprotryne	SCH_3	H	$CH(CH_3)_2$	N_2		3.69	26
	20	Methoprotryne	SCH_3	H	$CH(CH_3)_2$	H	$(CH_2)_3OCH_3$	-	51

lar properties. Limiting factor in the selection of triazines to include in
the calibration set was the availability of data for such, or other, common
descriptors. Collected data were analysed by PLS, which is a method designed
in such a way as to provide linear models of the dependence of the activity
from "latent variables"; the latter being linear combinations (principal component models) of the descriptors, developed under the constraint to maximize
the relationships with the activity. Among the PLS models of different dimensions provided by the analysis of our data, the most significant for the purpose of predictions of activity was the three dimensional model obtained from
a training set of 17 triazines (n.s 7 and 19 were omitted due to ill-defined
pI_{50} and lack of a number of descriptors, respectively). This model could explain 70% of the variance of the activity, and allowed to calculate the pI_{50}
values with a mean residual of 0.39 (in pI_{50} units). The plot of the experi-

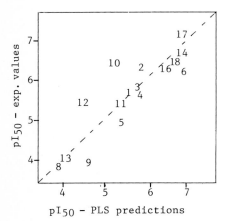

pI_{50} – PLS predictions

mental _versus_ calculated values (Fig.)
shows that the largest residuals belong
to two OCH_3-triazines, n.s 10 and 12,
and that, with few exceptions, calculated
and experimental values overlap satisfac-
torly. Tha PLS analysis also showed that
only 4 out of the 8 molecular properties
considered, were relevant to the descip-
tion of the end-point; these were the
chromatographic indices, accounting for
the partitioning of chemicals between

phases of different polarities, and, to a lesser extent, the water solubility.
QSAR models generated by PLS techniques may permit satisfactory predictions of
activity for untested chemicals, provided that they really belong to the same
class used for calibrating the model; appropriate procedure are available to
check, prior to predictions, whether this requirement is met (Dunn W.J. et al., 1984)

Qualitative criteria for estimates of acute toxic effects on _Daphnia_.
The measured EC_{50} show that the toxicity of triazines to _Daphnia_ lies within
a small range of values. However, the examined triazines are not necessarily
representative of this class of herbicides, and values outside this range may
well exist. Proper use of available data, by mean of structure-activity analysis criteria, may provide this information.

The PLS algorithm did not identify a satisfactory dependence of EC_{50} from the descriptors previously used: this may be attributed to the nature of the observed endpoint (immobilisation of Daphnia) which may be affected by uncontrolled factors, or to the inadequacy of the descriptors used to model such endpoint. Among alternative criteria of analysis of data we have empirically tried to separately examine the effects produced on EC_{50} by variyng the three X-substituents(within series of triazines with equal N,N-substitution) and those produced by varying the R_1-R_4 substituents(within series of triazines with equal X-substituent). The results of this analysis, which was only applicable to triazines sharing appropriate common substituents, are summarized in TABLE II.

TABLE II — Structural components and EC_{50} values of selected series of triazines

R_1R_2/R_3R_4		H,R_2/H,R_4			H,R_2/R_3,R_4
X		H,Et/H,iPr	H,iPr/H,iPr	H,Et/H,tBu	H,Et/Et,Et
		a	b	c	d
OCH$_3$	x	– atratone	38 prometone	44 terbutone	–
Cl	y	> 39 atrazine	11 propazine	> 5 terbutylazine	6 trietazine
SCH$_3$	z	40 ametryne	9.7 prometryne	7.1 terbutryne	–

Considering the values in the columns a, b and c, the SCH$_3$ and Cl groups appear to produce equivalent effects, and more severe than the OCH$_3$ group (column b and c); consideration of the values in the rows x,y and z suggests a trend of increasing toxicity with increasing bulkiness and number of N,N-substituents different from hydrogen (for example, the matrix element d/y is the most toxic chemical examined). Along these lines one should expect that the untested atratone (matrix element a/x) should be less toxic than all examined compounds, whereas triazines with all R_1-R_4 alkyl groups different from H should be the most toxic.

In conclusion, availability of homogeneous data sets for homogeneous chemicals is a necessary prerequisite for structure-activity analysis; this may be quantitative in presence of relevant descriptors, possibly multivariate, to be analyzed by adequate techniques; may otherwise, in absence of relevant descriptors, be qualitative and provide indications on trends of toxicity/activity in terms of structural characteristics of chemicals in a given class.

REFERENCES

Cesareo D., Tosato M.L. a,d Clementi S. (1986) Modelling phytotoxicity of herbicidal triazines by molecular properties. Environ.Toxicol. & Chem., in press.

Dunn, III, W.J., Wold S., Edlund U., Hellberg S. and Gasteiger J. (1984) Multivariate QSAR between data from a battery of biological tests and an ensemble of structure descriptors: the PLS method. Quant.Struct.Act.Relat. 3,131-137.

Ebert R. and Dumford S.W. (1976) Effects of triazines herbicides on the physiology of plants. Residue Review, vol. 65, 1-19.

Marchini S., Passerini L., Cesareo D. and Tosato M L. (1986) Acute toxicity of herbicidal triazines to Daphnia magna.Ecotoxicol. & Environ. Saf.,in press.

Decrements in Retrograde Axonal Transport Precede Onset of Nerve Fiber Degeneration in Organophosphate Neuropathy

A. Moretto*, M. Lotti**, M.I. Sabri, P.S. Spencer
Albert Einstein College of Medicine
1300 Morris Park Avenue
Bronx, New York, 10461, USA

A single dose of some organophosphorous (OP) esters induces a central-peripheral distal axonopathy known as OP-induced delayed polyneuropathy, OPIDP (Johnson, 1982). Phosphinates, carbamates or sulphonates, such as phenylmethanesulphonyl fluoride, given to animals prior to administration of an OP which causes OPIDP, prevent the onset of OPIDP (Johnson, 1982). The biochemical mechanism of initiation of OPIDP is inhibition and 'aging' (within hours of dosing) of Neuropathy Target Esterase (NTE) by a neuropathic OP (Moretto and Johnson, in this volume). Protective compounds inhibit NTE, but the inhibited enzyme does not 'age'. The threshold for initiation of OPIDP is 70-80% of inhibition/'aging' of NTE. Therefore, if 30% or more of NTE is occupied by a protective compound, a subsequent challenge dose of a neuropathic OP fails to cause neuropathy. It appears that the critical OP-NTE interaction occurs in the nerve fiber (presumably in the axon), rather than in the cell body (Lotti et al., 1987). However, the events that intervene between the two-step initiation mechanism and the onset of OPIDP approximately two weeks later are unknown. We studied the effect of a single dose of a neurotoxic OP (di-n-butyl-2,2-dichlorovinyl phosphate) on retrograde axonal transport of ^{125}I-tetanus toxin (Stoeckel et al., 1975) in peripheral motor and sensory axons of the hen in the period between the initiation and the clinical/morphological expression of OPIDP.

Materials and Method

Di-n-butyl-2,2-dichlorovinyl phosphate (DBDCVP) and Mipafox (N,N-diisopropyl phosphorodiamidofluoridate) were kindly supplied by Dr. R.J. Richardson (Ann Arbor, Mi). Paraoxon (diethyl p-nitrophenyl phosphate) and phenylmethanesulphonyl fluoride (PMSF) were purchased from Sigma Chem. Co. (St. Louis, Mo);

* present address: Istituto di Medicina del Lavoro, Universita' di Padova, via Facciolati 71, 35127 Padova, ITALY.
** Istituto di Medicina del Lavoro, Universita' di Padova, via Facciolati 71, 35127 Padova, ITALY.

NATO ASI Series, Vol. H13
Toxicology of Pesticides: Experimental, Clinical and Regulatory Aspects. Edited by L. G. Costa et al.
© Springer-Verlag Berlin Heidelberg 1987

the former was purified according to Johnson (1977). Phenyl valerate was synthesized and purified as described by Johnson (1977). Groups of adult hens (1.5-1.9 kg body weight) were treated as follows: a) DBDCVP, 0.75 mg/kg; b) PMSF, 30 mg/kg; c) Paraoxon, 0.2 mg/kg; d) PMSF, 30 mg/kg plus DBDCVP, 1 mg/kg 24 h thereafter. Compounds were dissolved in glycerol formal and injected subcutaneously. NTE and AChE activities in brain, spinal cord and sciatic nerve were determined as described (Ellman et al., 1961; Johnson, 1977; Caroldi and Lotti, 1982) 24 h after dosing. Walking performance was evaluated according to a 4-point scale (Johnson and Barnes, 1970) from day 7 after treatment until sacrifice (day 15-21). For morphological evaluation, peripheral nerves (sciatic and tibial) and spinal cord (cervical and lumbosacral) were excised from animals perfused with phosphate-buffered paraformaldehyde/glutaraldehyde (pH 7.4). Tissue samples were post-fixed in Dalton's chrome osmium, dehydrated stepwise and embedded in Polybed. One-micrometer sections were stained with toluidine blue and examined by light microscopy. Purified tetanus toxin, iodinated as described (Miller and Spencer, 1984), was unilaterally injected in the gastrocnemius muscle of hens (10 uCi in 5 ul). Ipsilateral and contralateral dorsal root ganglia of roots II, III, and IV of the plexus ischiaticum, and the corresponding sections of ventral spinal cord, were excised and their ^{125}I content determined with a Packard Multiprias gamma counter. Protein content was estimated in each tissue sample (Lowry et al., 1951). Accumulation of ^{125}I in ipsilateral tissues was considered an index of axonal transport:

$$\text{axonal transport} = \text{CPM/mg protein}_{ipsi} - \text{CPM/mg protein}_{contra}$$

To determine the effect on the rate of retrograde axonal transport of vehicle or DBDCVP (1 mg/kg) administration seven days earlier, animals were unilaterally injected in gastrocnemius muscle with ^{125}I-tetanus toxin and the position of the leading edge of transported radioactivity along the sciatic nerve determined after 2.5 and 6.5 h. The leading edge of transported material was considered to be the most proximal 5-mm segment of the ipsilateral sciatic nerve containing greater than 1% of the total radioactivity of the whole nerve (considered an estimate of nerve fiber uptake of radiolabel), calculated as:

$$\text{total radioactivity} = \sum_{1_i}^{\eta} (\text{CPM}_{ipsi} - \text{CPM}_{contra}).$$

The distance of the leading edge from the site of injection was plotted against the latency between ^{125}I-tetanus toxin injection and sacrifice. The slope obtained from linear regression analysis of the data was considered to represent the rate of retrograde transport of ^{125}I-tetanus toxin.

Results and Discussion

DBDCVP-treated animals showed the first clinical signs of OPIDP on day 10 and by day 15, functional deficit was fully developed (Table I). Morphological evaluation on day 21 revealed the characteristic pattern of axonal degeneration in dorso-lateral columns of cervical spinal cord, in ventral columns of lumbosacral spinal cord, and sciatic and tibial nerves. Animals treated with PMSF or paraoxon developed no neurological signs (table I). Birds which received PMSF 24 hours prior to a neurotoxic dose of DBDCVP failed to develop any sign of OPIDP.

	% activity in peripheral nerve*		clinical score
	NTE	AChE	
DBDCVP 0.75 mg/kg	14 ± 3 (7)	71 ± 7 (3)	3.5 ± 0.2 (6)
PMSF 30 mg/kg	18 ± 4 (4)	78,89	0 (2)
Paraoxon 0.2 mg/kg	118 ± 5 (4)	63 ± 3 (4)	0 (2)
PMSF 30 mg/kg plus DBDCVP 1.00 mg/kg	5,7	not done	0 (2)

Table I: Biochemical and clinical effects of treatment with different compounds. Percentage of activity was calculated from the activity of peripheral nerves obtained from control birds dosed with the vehicle alone on the same day. NTE activity (mean + SEM, uMoles/min/g of tissue) of 8 controls was 0.14 ± 0.01, AChE activity (mean + SEM, uMoles/min/g of tissue) of 5 controls was 1.5 ± 0.2. Clinical score was assessed on day 15. All data are expressed as means + SEM; ( ) = number of animals.
* % activity in brain and spinal cord was comparable to that in peripheral nerve.

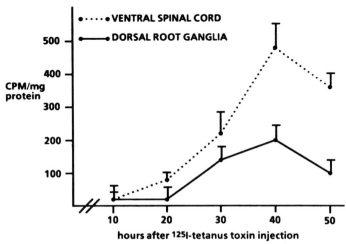

Figure I: Time-course of ^{125}I accumulation in ventral spinal cord and dorsal root ganglia after ^{125}I-tetanus toxin injection in hen gastrocnemius muscle. Values are mean + SEM, n=5.

The time-course of accumulation of ^{125}I in ipsilateral ventral spinal cord and dorsal root ganglia after unilateral injection of ^{125}I-tetanus toxin in hen gastrocnemius muscle is shown in Figure 1. Retrograde axonal transport of ^{125}I-tetanus toxin in treated animals was measured by assessing ^{125}I accumulation in ipsilateral tissues 36 h after ^{125}I-tetanus toxin injection. Temporally progressive decrements in ^{125}I accumulation in both ventral spinal cord (Fig. 2a) and dorsal root ganglia (Fig. 2b) were induced by a single dose of DBDCVP. Maximal effect on ^{125}I accumulation was reached at day 7, prior to the appearance of myelinated fiber degeneration in sciatic and tibial nerves. The onset of clinical/morphological evidence of OPIDP was not associated with a further decrease of ^{125}I accumulation when measured on day 21.

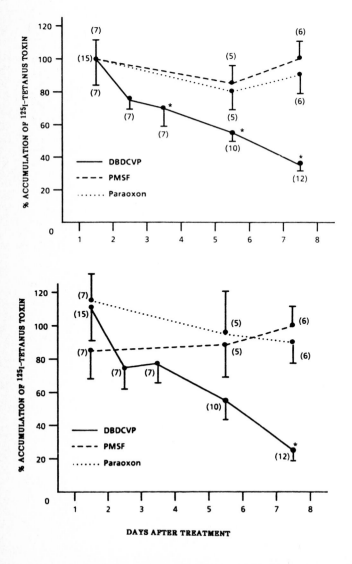

Figure 2: Effect of different treatments on ^{125}I accumulation in ventral spinal cord (a) and dorsal root ganglia (b). ^{125}I content was determined 36 hours after unilateral ^{125}I-tetanus toxin injection in gastrocnemius muscle. Percentage of ^{125}I accumulation was calculated from values obtained from control birds dosed the same day with vehicle alone. In control birds, ^{125}I accumulation in ventral spinal cord was 298 ± 15 CPM/mg of protein (mean ± SEM, n = 49) and in dorsal root ganglia 99 ± 7 (mean ± SEM, n = 48). Values are means ± SEM, ( ) = number of animals.

DAYS AFTER TREATMENT

treatment	rate(mm/h)	total radioactivity (CPM)
vehicle	5.50 ± 2.37(9)	26,395 ± 5,525(5)
DBDCVP 1 mg/kg	1.69 ± 1.48(9)	27,353 ± 6,830(4)

Table 2: Effect of DBDCVP on the rate of retrograde axonal transport of ^{125}I-tetanus toxin and total nerve-associated radioactivity. ^{125}I-tetanus toxin was injected in gastrocnemius muscle 7 days after DBDCVP or vehicle treatment. Total nerve-associated ^{125}I was calculated 6.5 hours after radiolabel injection. Data are means ± SEM, ( ) = number of animals.

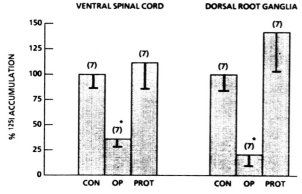

Figure 3: Protective effect of PMSF against DBDCVP-induced reduction of ^{125}I-tetanus toxin accumulation in ventral spinal cord and dorsal root ganglia. ^{125}I content was determined 36 hours after unilateral ^{125}I-tetanus toxin injection in gastrocnemius muscle and 7 days after vehicle or DBDCVP treatment. Protected animals received PMSF 24 hours before DBDCVP treatment. Values are means ± SEM. ( ) = number of animals.
CON = control;
OP = DBDCVP-treated;
PROT = protected.

A single dose of DBDCVP (1 mg/kg) also slowed the rate of retrograde axonal transport (Table 2) when measured 7 days after treatment. Total nerve radioactivity in DBDCVP-treated animals was not significantly different from that of vehicle-treated animals (Table 2) suggesting there was no significant effect on uptake of ^{125}I-tetanus toxin. A single dose of PMSF or paraoxon had no significant effect on ^{125}I accumulation in either dorsal root ganglia or ventral spinal cord (Figs. 2 and 3). PMSF pretreatment completely abolished the effect of DBDCVP on retrograde transport of ^{125}I-tetanus toxin (Fig. 4). This finding, together with the lack of changes in the negative control group treated with paraoxon, demonstrates that the reduction of retrograde axonal transport is intimately related to the development of OPIDP. The observed decrements in fast retrograde transport contrast with the reported absence of significant abnormalities in peripheral nerve anterograde axonal transport in OPIDP (Pleasure et al., 1969; James and Austin, 1970; Bradley and Williams, 1973), although a defect of anterograde transport in the most distal parts of the axon cannot be excluded. A reduction of retrograde axonal transport in the presence of normal anterograde transport could explain the axoplasmic accumulation of large vesicular and membranous structures which appear prior to axonal degeneration (Prineas, 1969; Bouldin and Cavanagh, 1979).

In conclusion, when a systemically delivered dose of OP penetrates nervous tissues and induces a supra-threshold inhibition/aging of axonal NTE, deficits in axonal transport accrue with time, reach a peak within 7 days and, thereafter, long and large-diameter fibers undergo distal degeneration and neurological deficits appear. Both the effect on retrograde axonal transport and the development of OPIDP are prevented by the organosulphonylation of the target protein, NTE, by PMSF. Thus, it is apparent that reduction of retrograde axonal transport is correlated with the structural modification ('aging') of the target protein NTE in the axon, and is not a consequence of major disruptive changes in the axon.

(Supported by OH 2065, NS 19611, Italian Ministry of Education, CNR 85.00581.56)

Bouldin TW, Cavanagh JB (1979) Organophosphorus neuropathy. II. A fine ultra-structural study of the early stages of axonal degeneration. Am J Path 94:253-270.

Bradley WG, Williams MH (1973) Axoplasmic flow in axonal neuropathies. I. Axoplasmic flow in cats with toxic neuropathies. Brain 96:235-246.

Caroldi S, Lotti M (1982) Neurotoxic esterase in peripheral nerve: assay, inhibition and rate of resynthesis. Toxicol Appl Pharmacol 62:498-501.

Ellman GL, Courtney DK, Andres V Jr, Featherstone RM (1961) A new and rapid colorimetric determination of acetylcholinesterase activity. Biochem Pharmacol 7:88-95.

James KAC, Austin L (1970) The effect of DFP on axonal transport of protein in chicken sciatic nerve. Brain Res 18:192-194.

Johnson MK (1977) Improved assay of neurotoxic esterase for screening for delayed neurotoxicity potential. Arch Toxicol 37:113-115.

Johnson MK (1982) The target for initiation of delayed neurotoxicity by organo-phosphorus esters: biochemical studies and toxicological applications. Rev Biochem Toxicol 4:141-212.

Johnson MK, Barnes JM (1970) Age and the sensitivity of chicks to the delayed neurotoxic effects of some organophosphorus compounds. Biochem J 19:3045-3047.

Lotti M, Caroldi S, Moretto A, Johnson MK, Fish CJ, Gopinath C, Roberts NL (1987) Central-peripheral delayed neuropathy caused by diisopropyl phosphorofluoridate (DFP): segregation of peripheral nerve and spinal cord effects using biochemical, clinical and morphological criteria. Toxicol Appl Pharmacol (in press).

Lowry OH, Rosenbrough NJ, Farr AL, Randall RJ (1951) Protein measurement with Folin phenol reagent. J Biol Chem 193:265-275.

Miller MS Spencer PS (1984) Single doses of acrylamide reduce retrograde transport velocity. J Neurochem 43:1401-1408.

Pleasure DE, Mishler KC, Engel WK (1969) Axonal transport of protein in experimental neuropathies. Science 166:524-525.

Prineas J (1969) The pathogenesis of dying-back polyneuropathies. I. An ultrastructural study of experimental tri-ortho-cresyl phosphate intoxication in the cat. J Neuropathol Exp Neurol 28:571-597.

Stoeckel K, Schwab M, Thoenen H (1975) Comparison between retrograde axonal transport of nerve growth factor and tetanus toxin in motor, sensory and adrenergic neurons. Brain Res 99:1-16.

The lungs as a target for the toxicity of some organophosphorus compounds.

B. Nemery[1]
Unité de Toxicologie Industrielle et Médecine du Travail
Université Catholique de Louvain
Clos Chapelle-aux-Champs 30 – Bte 30.54
1200 Bruxelles – Belgium

Lung toxicity of trialkyl phosphorothioates.

During the toxicological investigations which followed the malathion poisoning episode in Pakistan (Baker et al., 1978), it was discovered that in addition to their ability to potentiate the cholinergic toxicity of malathion (Aldridge et al., 1979), some trimethyl and triethyl phosphorothioates produced a form of toxicity which was unusual for organophosphorus compounds, in that they caused lung damage leading to deaths in respiratory distress 3 to 5 days after the administration of a single, relatively small dose (rat oral LD50's generally around or below 100 mg/kg) (Verschoyle & Cabral, 1982). Trimethyl and triethyl phosphorothioates are potential impurities in dimethyl and diethyl thiono-phosphorus insecticides. They have been detected, along with other contaminants, in technical or formulated samples of several commercial insecticides, such as malathion, phenthoate, acephate, fenitrothion, and diazinon (cf. Aldridge & Nemery, 1984). The specific pulmonary toxicity of trialkyl phosphorothioates has been confirmed by other investigators, and its characteristic features have been reviewed (Aldridge & Nemery, 1984; Aldridge et al., 1985a).

---

[1] Most studies described herein were conducted at the Medical Research Council Toxicology Unit in Carshalton, Surrey, U.K. under the supervision of Dr. W.N. Aldridge.

NATO ASI Series, Vol. H13
Toxicology of Pesticides: Experimental, Clinical
and Regulatory Aspects. Edited by L. G. Costa et al.
© Springer-Verlag Berlin Heidelberg 1987

Mechanisms for the lung toxicity of trialkyl phosphorothioates.

Trialkyl phosphorothioate-induced lung toxicity is not due to cholinesterase inhibition and deaths do not simply result from bacterial bronchopneumonia (Nemery et al., 1986). The target cell population in the lungs has been identified as the alveolar type 1 cells (Dinsdale et al., 1982). Important structure-toxicity relationships have been found : for trimethyl and triethyl phosphorothioates to produce lung damage they must possess at least one thiolo-alkyl moiety (CH_3-S- or C_2H_5-S-) without possessing a thiono-sulphur (P=S). The presence of a thiono-sulphur not only abolishes the lung toxicity, but it confers a considerable protection against the lung toxicity of the pneumotoxic trialkyl phosphorothiates (Umetsu et al., 1981; Aldridge et al., 1985a).

In addition to being vulnerable to many inhaled chemicals, the lungs are known to be the target of toxicity for a number of bloodborne chemicals (Kehrer & Kacew, 1985; Smith & Nemery, 1986). Although the reasons for the susceptibility of the lungs are often unknown, a useful framework for the study of mechanisms of chemical-induced lung toxicity has been proposed by Boyd (1980).

i) A first mechanism, exemplified by the herbicide paraquat, is that in which a chemical is selectively accumulated by active uptake in some lung cells where it then undergoes a cyclic reduction-oxidation process with the generation of toxic oxygen species in excess of the cells' antioxidant defences. This mechanism is probably not directly applicable to the toxicity of trialkyl phosphorothioates because there is no evidence that these compounds are concentrated in the lungs (Aldridge et al., 1984) and because their cyclic reduction-oxidation is difficult to envisage on chemical grounds.

ii) A second mechanism, exemplified by the pyrrolizidine alkaloid monocrotaline, is that in which hepatic metabolism results in the release of toxic metabolites which affect the pulmonary vascular bed. This mechanism is also considered to be unlikely to explain the lung toxicity of trialkyl phosphorothioates, on the grounds of the outcomes of various pretreatments with drug metabolism inducers or inhibitors (Aldridge et al., 1984, 1985a).

iii) The third mechanism, namely metabolic activation by pulmonary drug metabolizing enzymes themselves, is therefore probably responsible for the susceptibility of the lung to trialkyl phosphorothioates.

In order to test the latter hypothesis, in vitro metabolic studies were carried out.

Rat lung and liver microsomes or slices were incubated with O,S,S-trimethyl phosphorodithioate [OSSMe, $(CH_3S)_2(CH_3O)P=O$], labelled with ^3H or ^{14}C on one of its CH_3-S- groups.

Endpoints measured were 1) protein bound radioactivity (PBR); and 2) diester metabolites produced, either O,S-dimethyl phosphorothioate $[(CH_3S)(CH_3O)(OH)P=O$, OSMe] or S,S-dimethyl phosphorodithioate $[(CH_3S)_2(OH)P=O$, SSMe], measured by scintillation counting after separation by thin layer chromatography.

In microsomes, PBR and production of OSMe had features of cytochrome P-450-dependent processes : NADPH dependence, inhibition by CO, piperonyl butoxide, SKF525A, metyrapone; stimulation in liver, but not lung microsomes from phenobarbital-induced rats. Incubation with glutathione prevented the binding of radioactivity without affecting OSMe production. The rate of the conversion of OSSMe to OSMe was higher (in relation to microsomal cytochrome P-450 content) in lung than in liver microsomes, and the Km for this reaction was 15-fold lower in lung (0.3 mM) than in liver microsomes. Only minimal amounts of SSMe, known to result from metabolism of OSSMe by cytosolic glutathione-S-transferases (Aldridge et al., 1985b), were produced in microsomes.

In slices, PBR and the ratio OSMe/SSMe were much higher in lung than in liver. This suggests that in the lung the predominant route of metabolism for OSSMe is the cytochrome-P-450-mediated oxidation to OSMe, whereas in the liver the predominant route is the glutathione-S-transferase-mediated transfer of an O-methyl group to GSH, yielding methyl-S-glutathione and SSMe. Both PBR and OSMe production were decreased in lung, but not in liver slices, when the rats had been pretreated with O,O,O-trimethyl phosphorothioate $[(CH_3O)_3P=S$, 2.5 mg/kg p.o., 2 hours previously] or with p-xylene (1 g/kg i.p., 5 hours previously). These pretreatment schedules have been shown to be non-toxic for the lungs, and to protect against the lung toxicity of OSSMe, probably by inhibiting pulmonary mixed function oxidase activity (Aldridge et al. 1985a, Verschoyle & Aldridge, 1987).

The results of these in vitro studies suggest that the susceptibility of the rat lung to OSSMe results from a high rate of oxidation of OSSMe to OSMe in some lung cells. This cytochrome P-450-mediated reaction possibly involves oxidation of a thiolo-sulphur yielding a reactive sulphur-containing intermediate.

Human relevance of trialkyl phosphorothioate-induced lung injury ?

So far, the pulmonary toxicity of trialkyl phosphorothioates is only a laboratory observation on experimental animals, mainly the rat, but also the mouse (Kehrer et al., 1986), and it is not known whether and at which doses the human lung would be susceptible to those compounds. There is no suggestion in the reports of poisoning by contaminated insecticides (Baker et al., 1978; Soliman et al., 1982) of any pulmonary involvement in the affected subjects.

There is also little evidence or information in the literature about possible direct toxic effects of organophosphorus pesticides on the lungs, despite the knowledge that the lungs may be an important site for their binding (cf. Ellin, 1982) or their metabolism (Poore & Neal, 1972), despite the fact that exposure by inhalation may occur in occupational activities, and despite the fact that respiratory distress is commonly found in acute human poisoning with organophosphate insecticides (Haddad, 1983). The latter is generally attributed to the effects of cholinesterase inhibition leading to bronchospasm and excessive tracheobronchial secretions, respiratory muscle paralysis and depression of the respiratory centre. A possible complication of organophosphate insecticide poisoning is acute pulmonary oedema (Bledsoe & Seymour, 1972; Kass et al., 1978; Taveira da Silva, 1983); reportedly, organophosphate insecticide poisoning is by far the commonest cause of pulmonary oedema encountered in India (Wadia &

Sadagopan, 1973). However, the pathogenesis of such organophos-phate-induced pulmonary oedema is not clear. The acute and rapidly resolving forms, as found in acute parathion poisoning, have been attributed to muscarinic effects and decreased cardiac output, or to transient cerebral dysfunction and hypoxia, mechanisms which are also involved in narcotic- or barbiturate-induced oedema. Delayed pulmonary oedema or pneumonitis may also be observed in organophosphate poisoning, e.g. following diazinon (Namba et al., 1971) or parathion (Willems et al., 1971), but this seems to be usually attributed to aspiration of gastric contents or chemical pneumonitis from the petroleum distillate carrier (Haddad, 1983). Nevertheless, in all these instances the possibility of a direct pulmonary toxicity of the organophosphorus compound should not be completely dismissed.

The possible contribution of pesticide-exposure to the respiratory morbidity of farm workers is sometimes suggested, but this is not well-documented and pesticides other than organophosphorus compounds are usually implicated (cf. Seaton, 1984). The existence of a "pesticide lung" in fruit-growers and farmers during the spraying season (Lings, 1982) rests on anecdotal clinical suspicion and unconvincing epidemio-logical data, obtained without regard to the type of pesticides used.

The absence of convincing documentation of human lung disease directly attributable to organophosphorus insecticides, should not lead to exclude the need for a more thorough investigation of possible detrimental effects of these compounds on the respiratory system. More particularly, the direct human relevance of the pulmonary damage caused by trialkyl phosphorothioates in experimental animals remains an entirely open question, which could be partly answered by extending to human tissues the in vitro studies carried out on animal tissues. Meanwhile, the indirect relevance of studying the mechanisms of the lung toxicity of trialkyl phosphorothioates is that these compounds provide excellent model compounds for investigating the determinants of the lungs' susceptibility to chemical-induced damage, and that these studies have set the scene for the investigation of a hitherto somewhat neglected area in pesticide toxicology, namely the pneumotoxicology.

References

Aldridge WN, Nemery B (1984) Toxicology of trialkylphosphorothioates with particular reference to lung toxicity. Fundam Appl Toxicol 4:S215–S223.

Aldridge WN, Miles JM, Mount DL, Verschoyle RD (1979) The toxicological properties of impurities in malathion. Arch Toxicol 42:95–106.

Aldridge WN, Verschoyle RD, Peal JA (1984) O,S,S-trimethyl phosphorodithioate and O,O,S-triethyl phosphorothioate : pharmacokinetics in rats and effects of pretreatment with compounds affecting the drug processing systems. Pestic Biochem Physiol 21:265–274.

Aldridge WN, Dinsdale D, Nemery B, Verschoyle RD (1985a) Some aspects of the toxicology of trimethyl and triethyl phosphorothioates. Fundam Appl Toxicol 5:S47–S60.

Aldridge WN, Grasdalen H, Aarstad K, Nørkov T (1985b) Trialkyl phosphorothioates and glutathione S-transferase. Chem Biol Interact 54:243–256.

Baker EL, Zack M, Miles JW, Alderman L, Warren McW, Dobbin RD, Miller S, Teeters WR (1978) Epidemic malathion poisoning in Pakistan malaria workers. Lancet i:31–33.

Bledsoe FH, Seymour EQ (1972) Acute pulmonary edema associated with parathion poisoning. Radiology 103:53–56.

Boyd MR (1980) Biochemical mechanisms in chemical-induced lung injury : role of metabolic activation. CRC Crit Rev Toxicol 7:103–176.

Dinsdale D, Verschoyle RD, Cabral JRP (1982) Cellular responses to trialkylphosphorothioate-induced injury in rat lung. Arch Toxicol 51:79–89.

Ellin RI (1982) Anomalies in theories and therapy of intoxication by potent organophosphorus anticholinesterase compounds. Gen Pharmacol 13:457–466.

Haddad LM (1983) The organophosphate insecticides. In : Haddad LM, Winchester JF (eds) Clinical management of poisoning and drug overdose. WB Saunders Cy, Philadelphia, London : 704–710.

Kass JB, Khamapirad T, Wagner ML (1978) Pulmonary edema following skin absorption of organophosphate insecticide. Pediatr Radiol 7:113–114.

Kehrer JP, Kacew S (1985) Systematically applied chemicals that damage lung tissue. Toxicology 35:251–293.

Kehrer JP, Klein-Szanto AJP, Thurston DE, Lindenschmidt RC, Witschi HR (1986) O,S,S-trimethyl phosphorodithioate-induced lung damage in rats and mice. Toxicol Appl Pharmacol 84:480–492.

Lings S (1982) Pesticide lung : a pilot investigation of fruit-growers and farmers during the spraying season. Br J Ind Med 39:370–376.

Namba T, Nolte C, Jacrel J (1971) Poisoning due to organophosphate insecticides. Am J Med 50:475–492.

Nemery B, Tucker DK, Sparrow S (1986) A germ-free status does not protect from the lethal effects of acute lung damage caused by O,S,S-trimethyl phosphorodithioate. Toxicol Lett 32:153–162.

Poore PE, Neal RA (1972) Evidence for extrahepatic metabolism of parathion. Toxicol Appl Pharmacol 23:759–768.

Seaton A (1984) The breathless farmworker. Br Med J 288:1940–1941.

Smith LL, Nemery B (1986) The lung as a target organ for toxicity. In : Cohen GM (ed) Target organ toxicity. CRC Press, Boca Raton, Florida, In press.

Soliman SA, Sovocool GW, Curley A, Ahmed NS, El-Fiki S, El-Sebae AK (1982) Two acute human poisoning cases resulting from exposure to diazinon transformation products in Egypt. Arch Environ Health 37:207–212.

Taveira da Silva AM (1983) Principles of respiratory therapy. The shock lung and respiratory insufficiency/arrest. In : Haddad LM, Winchester JF (eds) Clinical Management of poisoning and drug overdose. Saunders Cy, Philadelphia, London :198–220

Umetsu N, Mallipudi NM, Toia RF, March RB, Fukuto TR (1981) Toxicological properties of phosphorothioate and related esters present as impurities in technical organophosphorus insecticides. J Toxicol Environ Health 7:481–497.

Verschoyle RD, Aldridge WN (1987) The interaction between phosphorothionate insecticides, pneumotoxic trialkyl phosphorothiolates and effects on lung 7-ethoxycoumarin-O-deethylase activity. Arch Toxicol (accepted).

Verschoyle RD, Cabral JRP (1982) Investigation of the acute toxicity of some trimethyl and triethyl phosphorothioates with particular reference to those causing lung damage. Arch Toxicol 51:221–231.

Wadia RS, Sadagopan C (1973) Organophosphates and pulmonary edema. N Engl J Med 104:289.

Willems J, Vermeire P, Rolly G (1971) Some observations on severe human poisonings with organophosphate pesticides. Arch Toxicol 28:182–191.

Behavioral and Biochemical Effects of Early Postnatal Parathion Exposure in the Rat.

C.R. Stamper, W. Balduini, S.D. Murphy and L.G. Costa
Department of Environmental Health, SC-34
University of Washington
Seattle, WA 98195
USA

The organophosphate insecticide, parathion, is a potent inhibitor of acetylcholinesterase both in the central and peripheral nervous systems. Acute exposure to this pesticide can cause a number of physical symptoms which are ameliorated as tolerance develops (Costa et al., 1982). There is, however, evidence of continuing behavioral deficits in humans following chronic exposure (Clark, 1971). These effects include impairment in concentration, decreased cognitive abilities, memory impairment, depression and irritability (Levin and Rodnitsky, 1976).

One of the major concerns of neurobehavioral toxicologists is the effect of exposure to toxins during pregnancy and early childhood. Exposure to parathion in utero has been shown to cause gross malformations in the fetus (Khera et al., 1969). However, these occurred at doses which caused significant toxicity to the rat dam. In general, as a class, organophosphates are not considered teratogens (Robens, 1969). Whereas moderate doses of parathion have not been shown to cause gross defects following prenatal exposure, they have been shown to cause long lasting behavioral effects in rats including delays in reflex development (Spyker and Avery, 1977), alterations in aggressive behavior, locomotor activity and learning tasks (Richardson et al., 1972).

To better extrapolate from animal models to humans, it is important to consider the stage of neural development in the experimental animal relative to humans. The postnatal brain growth spurt in the rat occurs from days 4-25 and corresponds to a time from mid gestation through the 4th year of life in the human (Dobbing, 1974). Critical events, including functional changes in a number of neurotransmitter systems , occur at this time. In the rat brain, the cholinergic system is not fully functional at birth, but reaches the adult level between one and two months (Coyle and Yamamura, 1976). Changes of various neurotransmitter levels during brain differentiation have been shown to cause persistant alterations of behavioral patterns in development and adulthood (Dorner, 1976).

NATO ASI Series, Vol. H13
Toxicology of Pesticides: Experimental, Clinical
and Regulatory Aspects. Edited by L. G. Costa et al.
© Springer-Verlag Berlin Heidelberg 1987

The purpose of the present study was to test the hypothesis that exposure to parathion during the postnatal, preweanling period (a time critical to brain and behavioral development) could cause deficits in the cholinergic system beyond the time of exposure.

METHODS

Male Long-Evans hooded rat pups from 20 litters were matched by weight and assigned to groups on postnatal day 5. Each litter consisted of 8-10 pups. Following preliminary studies in this laboratory to determine the acute effects of parathion in the neonatal rat, two doses were chosen for the present study: 1.9 mg/kg/day (50% of Day 5 LD_{50}) and 1.3 mg/kg/day (33% of Day 5 LD_{50}). The animals received subcutaneous injections of the high dose (n=29), low dose (n=30) or corn oil vehicle (n=29) from postnatal days 5-20.

Each day, the general condition and development of the animals was assessed. Daily body weights and day of incisor eruption and eye opening were recorded. A battery of reflex tests including a righting test, negative geotaxis, free-fall righting and cliff avoidance were conducted each day prior to injections (Almli and Fisher, 1977).

On day 21, 24 hours after the last injection, the brains were removed from a number of the animals in each group for analysis of [^{3}H]QNB binding, acetylcholinesterase (AChE) activity and protein levels in the cerebral cortex. Biochemical assessments were also performed on a group of animals exposed to 2.6 mg/kg/day of parathion from days 5-20. The remaining animals were weaned and housed for behavioral testing and growth assessment until postnatal day 40.

To assess spontaneous motor activity in a novel environment, each animal was tested in an automated open field apparatus on days 17 and 24. The number of vertical and horizontal movements were recorded each minute for 5 minutes. On postnatal day 24, each animal was tested for neuromuscular coordination and endurance on an accelerating Rota-rod treadmill. The length of time on the treadmill and the maximum speed attained were recorded for each rat.

Because spatial memory tasks in the rat have been shown to be mediated by the cholinergic system (Olton, 1983), a number of tests of spatial memory were conducted. On postnatal day 24, each animal was tested for spontaneous alternation in a T-maze. Spontaneous alternation refers to the tendency of the animal to enter opposite goal boxes on consecutive unrewarded trials in a T-maze (Douglas, 1966). A number of researchers have demonstrated that this sturdy, consistent behavior in rats can be disrupted by administration of the anticholinergic drug, scopolamine (Kokkinidis and Anisman, 1976).

For finer assessment of spatial memory deficits in these rats, we used an eight arm radial arm maze to differentiate between deficits in short term (working memory) and long term (reference memory)(Beninger et al., 1985). For this series of tests, arms 1-4 of the maze were baited with reward marshmallows. Arms 5-8 remained unbaited. Following a period of habituation and training, the animals were tested in the maze on days 37 and 38. Each animal was placed in the maze and allowed to remain until all eight arms had been entered. The length of time to accomplish this task and the individual arms entered were recorded. The dependent measures calculated from these data included the total time spent in the maze, the total number of errors in the maze (the number of arms entered greater than 8), the number of errors in working memory (returns to a baited arm of the maze), the number of reference memory errors (returns to a non-baited arm of the maze) and the percentage of time travelled in one given direction.

RESULTS AND DISCUSSION

The high dose animals exhibited mild symptoms of parathion toxicity including trembling and diarrhea for the first days of drug exposure, which gradually decreased over the course of the experiment. The low dose animals exhibited few, if any, symptoms. An analysis of variance revealed significant differences in the growth rate among the groups throughout preweanling development, $F(2,15)=24.4$, $p<.001$. On day 20, the last day of parathion exposure, the high and low dose group animals weighed 82% and 86% of controls, respectively. There were no significant differences among the groups in eye opening or incisor eruption. Analysis of reflex behavioral development showed differences on only one of the four tests, the cliff avoidance task, $F(2,46)=6.53$, $p<.003$. The mean latency to cliff avoid for the control group was 1.8 sec ±0.17, for the low dose group was 2.6 sec ±0.3 and for the high dose group was 3.2 sec ±0.4.

Significant treatment effects ($F(3,14)=9.5$, $p<.001$) and a significant unique linear trend ($F(1,14)=22.9$, $p<.003$) in an Analysis of Variance revealed there were significant dose dependent decreases in acetylcholinesterase activity following parathion exposure. There were significant dose dependent reductions in the density of muscarinic receptors (B_{max}) on day 21 (Treatment effect, $F(3,14)=7.4$, $p<.003$ and linear trend, $F(1,14)=12.9$, $p<.003$.) There were no significant differences among the groups in K_d (receptor affinity) or protein levels.

Following parathion exposure, there were no differences among the groups in the postweanling growth rate, in spontaneous activity in the open field or in performance on the rotarod treadmill. There were significant, dose related differences in the rate of spontaneous alternation in the T-maze on postnatal day 24 (Jonckheere Test, $Z=2.95$, $p<.0025$).

There were no significant differences among the groups in the total time required to perform the radial arm maze task, in the total number of errors in the maze or in the number of errors in reference memory. There were, however, significant, dose dependent deficits in working memory, $F(2,62)=3.3$, $p<.04$. In addition, there were significant differences in the directional pattern of maze exploration among the groups. The control rats turned in one direction a greater percentage of time than the rats in either treated group, $F(2,62)=3.6$, $p<.03$.

Throughout the course of parathion exposure, the animals appeared essentially normal. Developmental milestones were not delayed and there were deficits on only one of the four reflex measures in parathion exposed animals. There were essentially no differences on general, non specific behavioral indices either at the time of parathion exposure or following the end of the treatment. However, in tests specific for and sensitive to spatial memory impairment, the parathion exposed rats performed less well than the control group rats. These deficits were not due to motivational factors as the length of time in the radial arm maze and the activity levels in the open field did not differ. These deficits were also not due to undernutrition, as the level of growth retardation observed in this study was not sufficient to induce behavioral changes following body weight recovery (Leathwood, 1978). In addition, the altered performance cannot be attributed to problems of neuromuscular performance as there were no differences among the groups in rotarod performance. Exposure to parathion during the postnatal period resulted in dose dependent decreases in AChE activity and muscarinic receptor binding. These biochemical alterations would explain the differences in the performance on the spatial memory measures designed to specifically test cholinergic functioning.

In conclusion, parathion, at doses which did not cause overt symptoms of toxicity also did not cause lasting growth deficits or substantial alterations in most reflex behaviors; however, impaired performance in a number of behaviors of spatial memory related to the cholinergic system were observed. The extent of these deficits, and the time course of recovery remain to be studied.

ACKNOWLEDGEMENTS

This study was supported by a NATO grant for International Collaboration in Research and by grants ES-7032 and ES-03424 from NIEHS.

REFERENCES

Almli, C.R. and R.S. Fisher. Infant rats: sensorimotor ontogeny and effects of substantia nigra destruction. Brain Res.Bull.2(6):425-459, 1977.

Beninger, R.J. and B.A. Wirsching. Effects of altered cholinergic function on working and reference memory in the rat. Can. J. Physiol. Pharmacol. 64:376-382, 1985.

Clark, G. Organophosphate insecticides and behavior, a review. Aerospace Med. 42(7):735-740, 1971.

Costa, L.G., B.W. Schwab and S.D. Murphy. Tolerance to anticholinesterase compounds in mammals. Toxicology 25:79-97, 1982.

Coyle, J.T. and H.I. Yamamura. Neurochemical aspects of the ontogenesis of cholinergic neurons in the rat brain. Brain Res. 118:429, 1976.

Dobbing, J. The Later Development of the brain and its vulnerability. In: J. Davis and J. Dobbing, eds. Scientific Foundations of Pediatrics, 565-575, 1974.

Dorner, G. Hormones and brain differentiation. Elsevier, Amsterdam, 243-263, 1976.

Douglas, R. Cues for Spontaneous Alternation. J.Comp.Physiol.Psychol. 62(2):171-183, 1966.

Khera, K.S. and D.J. Clegg. Perinatal toxicity of pesticides. Canad.Med.Assoc.J. 100:167-172, 1969.

Kokkinidis, L. and H. Anisman. Dissociation of the effects of scopolamine and d-amphetamine on a spontaneous alternation task. Pharm.Biochem.Behav. 5(3):293-297, 1976.

Leathwood, P. Influence of early undernutrition on behavioral development and learning in rodents. In: G. Gottlieb, ed. Early Influences, 187-209, 1978.

Levin, H.S. and R.L. Rodnitzky. Behavioral effects of organophosphate pesticides in man. Clin.Toxicol. 9(3):391-405, 1976.

Olton, D. The use of animal models to evaluate the effects of neurotoxins on cognitive processes. Neurobehav.Toxicol.Teratol. 5(6):635-640, 1983.

Richardson, D.L., A.G. Karczmar and C.L. Scudder. Effects of pre-natal cholinergic drug treatment on post-natal behavior and brain chemistry in mice. Fed.Proc. 31:596, 1972.

Robens, J.F. Teratologic studies in carbaryl, diazinon, noria, disulfiram and thiram in small laboratory aninmals. Toxicol.Appl.Pharmacol. 15:152-163, 1969.

Spyker, J.M. and D.L. Avery. Neurobehavioral effects of prenatal exposure to the organophosphate diazinon in mice. J. Toxicol and Env. Health. 3:989-1002, 1977.

EFFECTS OF LINDANE ON CENTRAL NERVOUS SYSTEM: BEHAVIOURAL STUDIES

J.M. Tusell, C. Suñol, J. Llorens, E. Gelpí and * E. Rodríguez-Farré.
Dept. Neurochemistry. *Dept. Pharmacology and Toxicology. Consejo Superior
Investigaciones Científicas. Jorge Girona Salgado 18-26. 08034 Barcelona.

INTRODUCTION

Lindane (γ-hexachlorocyclohexane) is used as a general insecticide as
well as in the treatment of scabies and pediculosis.

Acute high doses of lindane act upon the mammal central nervous system
(CNS) producing a series of hyperexcitability signs, such as myoclonic jerks
and convulsions. The mechanisms of its neurotoxicity are far from being clarified (Woolley et al, 1985).

In this work we studied the relationship between lindane concentration in
brain and the initial convulsant response after administration of different
doses. We also presents the preliminary results using a residential maze to
study the behavioural effects of non convulsant doses of lindane.

MATERIALS AND METHODS

Animals and chemicals

Male Wistar rats (150-180 g) were purchased from Iffa Credo (Saint Germain-
sur-l'Arbresle). The animals were supplied standard diet pellets and water ad
libitum, and were maintained at 22+2ºC in a dark/light cycle of 12 hours for
at least 7 days before experimentation. Lindane (Merck, 99.5%) was checked for
chemical purity by gas chromatography, dissolved in olive oil and administered
by gavage (po) or by intraperitoneal injection (ip). Amphetamine and chlorpromazine dissolved in saline were given subcutaneously (sc).

Animal observation and locomotor activity determination

The responses selected as end-points for measuring the neurotoxic action
were both the first tonic seizure and the locomotor activity after administration of a single dose of lindane. For measuring the spontaneous locomotor activity the four residential mazes used were placed in a soundproof chamber.
Optical gate breaks were registered by a special electronic device (Panlab,
Barcelona) and transferred to a HP 85 computer through a BCD interface. The
sessions finish 4 hours after introducing the rats into the maze.

Determination of lindane

Lindane in brain and blood was extracted with hexane and determined in a

NATO ASI Series, Vol. H13
Toxicology of Pesticides: Experimental, Clinical
and Regulatory Aspects. Edited by L. G. Costa et al.
© Springer-Verlag Berlin Heidelberg 1987

gas chromatograph (Perkin Elmer 900) equipped with an Electron Capture Detector. Brain lindane levels in rats treated with convulsant doses were determined 40 min after dosing. In activity studies rats were sacrificed 15 mn after the maze session.

Analysis of data

The results are expressed as mean $\pm$ SEM. Linear correlations and regression between variables were fitted by the least-squares method. The area under the concentration-time curve (AUC) was calculated by the trapezoidal rule and by integration of the monoexponential equation from t= 168 hours to t=∞. The total maze activity measures were compared statistically using the one-sided Student's t-test and ANOVA.

RESULTS

Relationships between dose, concentration and convulsant response

Table I summarizes the results of the experiments conducted with the tonic seizure as end-point for determining the acute neurotoxic response of rats to different po doses of lindane.

The most important effect observed after lindane administration was a series of hyperexcitability signs with at least one myoclonic seizure eventually resulting in a tonic seizure. The incidence rate (number of animals having tonic seizures) was dose-dependent. There was a dose-related increase of lindane concentration in brain. A significant correlation for the linear relationship between brain concentration and dose was observed ($p < 0.01$). The incidence of response was linearly correlated ($p < 0.05$) with the log of lindane concentration in brain in the range of 12-80% of response for a single po administration.

Table I.- Convulsant effect of lindane

Dose	Number of animals	Lindane brain concentration	Incidence rate of convulsions
mg/kg	n	μg/g	%
60	16	4.2 $\pm$ 0.1	12.5
100	20	5.6 $\pm$ 0.4	60
150	10	6.5 $\pm$ 0.5	80

Kinetics of lindane in brain and blood

Figure 1 shows the parallelism between lindane kinetics in brain and blood after po and ip administration.

The results of the kinetic analysis were carried out considering the open one-compartment model as the best description for concentrations that appear to follow first-order changes. While the absorption half-life was similar for both routes of administration (~1 h), the elimination rate for brain (45 h vs 27 h) and blood (38 h vs 29 h) was slower for lindane ip injected than orally given. Also the AUC in brain and blood was higher for the po than the ip route. From the AUC it was calculated that lindane by ip route had a relative bioavailabilty of about 80% when compared with po route.

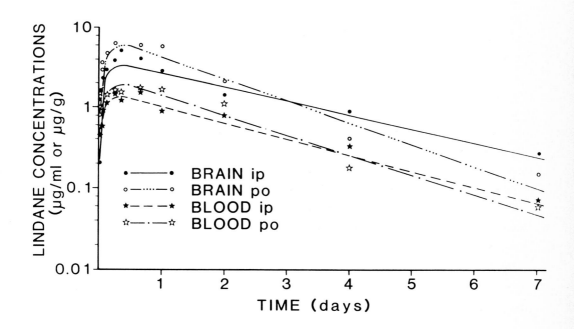

Figure 1.- Time-course of lindane concentrations in brain and blood after
          a sigle po or ip administration (60 mg/kg) at times ranging from
          5 min to 7 days. Each kinetics in brain or blood was established
          from determination in 42-48 animals.

Effect of lindane on spontaneous locomotor activity

Table II shows the results of the locomotor activity studies. There were not differences between saline, control and olive oil treated rats. The method was validated using amphetamine and chlorpromazine as reference

compounds, according EPA (1982).There was a significative increase between
the group of rats treated with amphetamine compared with saline group.
Lindane treated rats (30 mg/kg) also showed a significative increase in
activity compared with olive oil treated rats.

Table II.- Effect of lindane, vehicles and reference compounds on the total
activity (4-hours maze sessions)

Treatment	Dose	Total activity	Number of animals
	mg/kg	counts: $\bar{x}$ + SEM	n
Amphetamine	1	$1857.4 + 150.0^{a}$	8
Chlorpromazine	4	$174.8 + 12.9^{a}$	8
Saline	–	$914.8 + 102.7^{b}$	8
Control	–	$979.5 + 65.4^{b}$	8
Olive oil	–	$908.9 + 57.4^{b}$	15
Lindane	30	$1288.7 + 127.0^{c}$	9

a) differs from saline (p $<$ 0.001, Student's t-test)

b) n.s. (ANOVA)

c) differs from olive oil (p $<$ 0.01, Student's t-test)

Amphetamine, chlorpromazine and saline were administered sc, and lindane
and olive oil po 30 min before starting the maze session (10.00 am).
Control rats received no treatment.

DISCUSSION

The best correlation between exposure and poisoning signs (convulsions)
was related to the concentration of lindane in brain at onset of seizures,
being the magnitude of the convulsant response directly proportional to
the log of concentration of unchanged lindane in the target tissue. The
convulsant response was also dose related and in good agreement with the
linear dose-dependent concentration of the compound in brain. Development
of acute tolerance to convulsant agents such as lindane causes the disappearance of biological effects long before the elimination of the toxic
agent from the target organ (Ramzan and Levy 1985). In these cases the
toxicant half-life is poorly correlated with biological effects. Instead
the toxicological end-point defined in this work was taken as the onset
of the first tonic seizure. The smallest oral dose that produced tonic

seizures (60 mg/kg) did not cause any mortality after non lethal doses of the chemical.

The pharmacokinetic characteristics observed after ip dosing may account for the lower toxicity of this route observed previously by us as a consequence of the enhanced accumulation of lindane in peritoneal adipose tissue (unpublished data).

To determine changes in spontaneous locomotor activity elicited by low doses of neurotoxic chemicals, we have developed and automated and sensitive method based in a residential maze similar to that described by Elsner et al. (1979). Regarding the total activity, its sensitivity is similar to other maze devices, as demonstrated by amphetamine and chlorpromazine elicited values. Lindane at single doses of 30 mg/kg increased the total activity.

In conclusion, the actual amount of the toxicant in the effector target organ is an essential information if exposure conditions are to be succesfully correlated with toxic response parameters. The degree of CNS-stimulation would determine a large neurotoxicological spectrum of responses going from the initial hyperexcitability , starting at low brain levels of lindane to the convulsant seizures and also including a series of behavioural and neurological changes of increasing severity at intermediate concentrations.

REFERNCES

Elsner J, Looser R, Zbinden G (1979). Quantitative analysis of rat behavior in a residential maze. Neurobehav Toxicol 1 (Suppl 1) 163÷174

EPA (1982). Health effects test guidelines. EPA 560/6-82-001. National Technique Information Service, Springfield, Va.

Ramzan IM, Levy G (1985). Kinetics of drug action in disease states. XIV. Effect of infusion rate on pentylenetetrazol concentrations in brain and cerebrospinal fluid of rats at onset of convulsions. J Pharmacol Exp Ther 234:624-628

Woolley D, Zimmer L, Dodge D, Swanson R (1985). Effects of lindane-type insecticides in mammals: unsolved problems. Neurotoxicology 6:165-192

Subject Index

Acceptable Daily Intake (ADI)
128, 133, 139, 222
Acetylcholinesterase (AChE) 33,
201, 305
- aging 40
- reactivation 39
Acute toxicity 11-17, 27, 33,
154
- in man 197-206
- LD_{50} 23, 49, 50, 154-155
- SKF-525A effect 23
Adrenergic system 78-79
Agent Orange 161
Ah locus 96-97
Aldrin 4, 80, 96
Alimentary epidemics 273
Amitraz 78
Antioxidants 71
Arsenical pesticides 1, 163-164
ATPase inhibition 80
Atrazine 285
Avermectin B_{1a} (AVM) 83-84
Azinphos-methyl (see methyl
guthion)
Axonal transport 291

Bacillus Thuringiensis 241
Behavioural effect 85, 305, 311
Benomyl
- dermal toxicity 161
- reproductive toxicity 115
- teratogenicity 118
Biotechnology 241-252
Butyrylcholinesterase (see
pseudo - ChE)

Captan
- dermal toxicity 161
- teratogenicity 118
Carbamates 6
- carcinogenicity 140-142
- chemical structure 34

- clinical effects 202-203
- clinical treatment 203
- mechanism of action 33, 140
- metabolism 98-100
- residues 190-192
- toxicity of 33-47
Carbaryl 6
- carcinogenicity 142
- reproductive toxicity 113-117
Carbophenothion 190, 193, 275
Carcinogenicity 49, 56, 103-105,
125-145
- chlorphenoxy herbicides 56
- cutaneous 158-159
Cell injury 63-75
Central nervous system effect 36,
53
- insecticides and 77-91, 198-199
Chloracne 50, 157-158, 162
Chlordane 80, 150
- carcinogenicity 132-133
- reproductive function 114-115
Chlordimeform 78
Chlorfenvinphos 190
Chlorphenoxy herbicides 49-61
- clinical effects 204
- immunotoxicity 55
- metabolism 50
- neurologic effects 52
- reproductive toxicity 53
- teratogenicity 53
Cholinergic system 84-87
- inhibition of acetylcholinesterase
5-20, 23 33-39, 45, 104, 197, 279
- muscarinic manifestations
36, 198
- nicotinic manifestations 36, 198
Chronic toxicity 27, 197
Clinical evaluation 11-17,
210-212
Clinical management 197-206
Contact dermatitis 50, 156
Contaminant toxicity 50
Crop protection 4

NATO ASI Series H